# METHODS IN MOLECULAR BIOLOGY

*Series Editor*
**John M. Walker**
**School of Life and Medical Sciences**
**University of Hertfordshire**
**Hatfield, Hertfordshire, AL10 9AB, UK**

For further volumes:
http://www.springer.com/series/7651

# Clinical Applications of Mass Spectrometry in Drug Analysis

**Methods and Protocols**

Edited by

## Uttam Garg

*Department of Pathology & Lab Medicine, Children's Mercy Hospitals & Clinics, Kansas City, MO, USA*

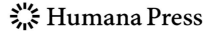 Humana Press

*Editor*
Uttam Garg
Department of Pathology & Lab Medicine
Children's Mercy Hospitals & Clinics
Kansas City, MO, USA

ISSN 1064-3745          ISSN 1940-6029   (electronic)
Methods in Molecular Biology
ISBN 978-1-4939-3251-1        ISBN 978-1-4939-3252-8   (eBook)
DOI 10.1007/978-1-4939-3252-8

Library of Congress Control Number: 2015956617

Springer New York Heidelberg Dordrecht London

Printed on acid-free paper

Humana Press is a brand of Springer
Springer Science+Business Media LLC New York is part of Springer Science+Business Media (www.springer.com)

# Dedication

This book is dedicated to my wife Jyotsna and daughters Megha and Mohini who are my inspiration to keep moving forward in life.

# Preface

Applications of mass spectrometry particularly in the fields of drugs and toxins analyses are growing exponentially. This is due to the facts that the newer instruments are much more user-friendly, less expensive, compact, and robust. Furthermore, the technology is versatile and generally does not require any special reagents such as antibodies to the drugs. Other advantages of mass spectrometry include increased sensitivity and specificity and multicomponent analysis. Using this technology hundreds to thousands of compounds can be screened in a single assay run. This volume *Clinical Applications of Mass Spectrometry in Drug Analysis* provides methods and protocols for a number of drugs and toxins in a stepwise manner. Brief introduction and clinical utility of the analytes along with useful notes to help the readers easily reproduce the protocols are also provided.

I am indebted to my colleagues who took time from their busy schedule to contribute to the contents of this book. I hope this book will be useful to the laboratorians who are already using mass spectrometry or thinking of bringing this technology to their laboratories.

*Kansas City, MO, USA*                                                                                      *Uttam Garg*

# Contents

# Contributors

SUSAN ABDEL-RAHMAN • *Department of Pediatrics, Children's Mercy Hospitals and Clinics, Kansas City, MO, USA*

BRIAN D. AHRENS • *UCLA Olympic Analytical Laboratory, Department of Pathology & Laboratory Medicine, David Geffen School of Medicine at UCLA, Los Angeles, CA, USA*

GEZA S. BODOR • *Pathology and Laboratory Medicine Service, VA Eastern Colorado Health Care System, Denver, CO, USA*

AUTUMN R. BREAUD • *Department of Pathology, The Johns Hopkins University School of Medicine, Baltimore, MD, USA*

ANTHONY W. BUTCH • *UCLA Olympic Analytical Laboratory, Department of Pathology & Laboratory Medicine, David Geffen School of Medicine at UCLA, Los Angeles, CA, USA*

DEAN C. CARLOW • *Department of Laboratory Medicine, Memorial Sloan Kettering Cancer Center, New York, NY, USA*

ALLISON B. CHAMBLISS • *Department of Pathology-Clinical Chemistry, Johns Hopkins University School of Medicine, Baltimore, MD, USA*

WILLIAM A. CLARKE • *Department of Pathology-Clinical Chemistry, Johns Hopkins University School of Medicine, Baltimore, MD, USA*

JENNIFER COLBY • *Department of Laboratory Medicine, San Francisco General Hospital, University of California, San Francisco, CA, USA*

CHRISTOPHER A. CRUTCHFIELD • *Department of Pathology, The Johns Hopkins University School of Medicine, Baltimore, MD, USA*

JIGNESH DALAL • *Department of Pediatrics, Children's Mercy Hospitals and Clinics, Kansas City, MO, USA*

DARLINGTON DANSO • *Department of Laboratory Medicine and Pathology, Mayo Clinic, Rochester, MN, USA*

BREHON DAVIS • *Department of Pathology and Laboratory Medicine, Children's Mercy Hospitals and Clinics, Kansas City, MO, USA*

SHUANG DENG • *Department of Pathology and Laboratory Medicine, Children's Mercy Hospitals and Clinics, Kansas City, MO, USA; Department of Pediatrics, Children's Mercy Hospitals and Clinics, Kansas City, MO, USA*

MARTIN FLEISHER • *Department of Laboratory Medicine, Memorial Sloan Kettering Cancer Center, New York, NY, USA*

CLINT FRAZEE • *Department of Pathology and Laboratory Medicine, Children's Mercy Hospitals and Clinics, Kansas City, MO, USA*

MELISSA FUNKE • *Department of Pathology and Laboratory Medicine, Children's Mercy Hospitals and Clinics, Kansas City, MO, USA*

ANNA K. FÜZÉRY • *Department of Pathology-Clinical Chemistry, Johns Hopkins University School of Medicine, Baltimore, MD, USA*

UTTAM GARG • *Department of Pathology and Laboratory Medicine, Children's Mercy Hospitals and Clinics, Kansas City, MO, USA*

SHANNON HAYMOND • *Department of Pathology, Northwestern University Feinberg School of Medicine, Chicago, IL, USA*

PAUL J. JANNETTO • *Department of Laboratory Medicine and Pathology, Mayo Clinic, Rochester, MN, USA*

KAMISHA L. JOHNSON-DAVIS • *Department of Pathology, University of Utah Health Sciences Center, School of Medicine, ARUP Institute for Clinical and Experimental Pathology, Salt Lake City, UT, USA*

BRIDGETTE JONES • *Department of Pediatrics, Children's Mercy Hospitals and Clinics, Kansas City, MO, USA*

GREGORY KEARNS • *Department of Pathology and Laboratory Medicine, Children's Mercy Hospitals and Clinics, Kansas City, MO, USA; Department of Pediatrics, Children's Mercy Hospitals and Clinics, Kansas City, MO, USA*

MICHAEL KISCOAN • *Department of Pathology and Laboratory Medicine, Children's Mercy Hospitals and Clinics, Kansas City, MO, USA*

YULIA KUCHEROVA • *UCLA Olympic Analytical Laboratory, Department of Pathology & Laboratory Medicine, David Geffen School of Medicine at UCLA, Los Angeles, CA, USA*

LORALIE J. LANGMAN • *Department of Laboratory Medicine and Pathology, Mayo Clinic, Rochester, MN, USA*

STEPHANIE J. MARIN • *ARUP Institute for Clinical and Experimental Pathology, Salt Lake City, UT, USA*

MARK A. MARZINKE • *Department of Pathology, The Johns Hopkins University School of Medicine, Baltimore, MD, USA*

GWENDOLYN A. McMILLIN • *ARUP Laboratories, Salt Lake City, UT, USA; Department of Pathology, University of Utah School of Medicine, Salt Lake City, UT, USA*

ALAN MILLER • *Department of Pathology, Northwestern University Feinberg School of Medicine, Chicago, IL, USA*

ALEJANDRO R. MOLINELLI • *Department of Pharmaceutical Sciences, St. Jude Children's Research Hospital, Memphis, TN, USA*

ADA MUNAR • *Department of Pathology and Laboratory Medicine, Children's Mercy Hospitals and Clinics, University of Missouri School of Medicine, Kansas City, MO, USA*

KAZUNORI MURATA • *Department of Laboratory Medicine, Memorial Sloan Kettering Cancer Center, New York, NY, USA*

JAIME H. NOGUEZ • *Department of Pathology, University Hospitals Case Medical Center, Cleveland, OH, USA*

JUDY PEAT • *Department of Pathology and Laboratory Medicine, Children's Mercy Hospitals and Clinics, Kansas City, MO, USA*

MELISSA S. PESSIN • *Department of Laboratory Medicine, Memorial Sloan Kettering Cancer Center, New York, NY, USA*

LAKSHMI V. RAMANATHAN • *Department of Laboratory Medicine, Memorial Sloan Kettering Cancer Center, New York, NY, USA*

BHEEMRAJ RAMOO • *Department of Pathology and Laboratory Medicine, Children's Mercy Hospitals and Clinics, Kansas City, MO, USA*

JAMES C. RITCHIE • *Department of Pathology and Laboratory Medicine, Emory University Hospital, Atlanta, GA, USA*

ENGER ROBERT • *Department of Laboratory Medicine and Pathology, Mayo Clinic, Rochester, MN, USA*

ALAN L. ROCKWOOD • *Department of Pathology, University of Utah School of Medicine, Salt Lake City, UT, USA*

CHARLES H. ROSE IV • *Department of Pharmaceutical Sciences, St. Jude Children's Research Hospital, Memphis, TN, USA*

GEOFFREY S. RULE • *Institute for Clinical and Experimental Pathology, ARUP Laboratories, Salt Lake City, UT, USA*

RYAN C. SCHOFIELD • *Department of Laboratory Medicine, Memorial Sloan Kettering Cancer Center, New York, NY, USA*

HENG SHI • *Department of Pathology and Laboratory Medicine, Children's Hospital of Philadelphia, Philadelphia, PA, USA*

MATTHEW H. SLAWSON • *ARUP Institute for Clinical and Experimental Pathology, Salt Lake City, UT, USA; ARUP Laboratories, Salt Lake City, UT, USA*

JUDY STONE • *Toxicology/Mass Spectrometry Laboratory, Center for Advanced Laboratory Medicine, University of California San Diego, San Diego, CA, USA*

GINA VESPA • *Department of Pathology, Northwestern University Feinberg School of Medicine, Chicago, IL, USA*

FAYE B. VICENTE • *Department of Pathology, Northwestern University Feinberg School of Medicine, Chicago, IL, USA*

ALAN H.B. WU • *Department of Laboratory Medicine, San Francisco General Hospital, University of California, San Francisco, CA, USA*

YAN VICTORIA ZHANG • *Department of Pathology and Laboratory Medicine, University of Rochester, Rochester, NY, USA*

# Chapter 1

# Mass Spectrometry in Clinical Laboratory: Applications in Therapeutic Drug Monitoring and Toxicology

## Uttam Garg and Yan Victoria Zhang

## Abstract

Mass spectrometry (MS) has been used in research and specialized clinical laboratories for decades as a very powerful technology to identify and quantify compounds. In recent years, application of MS in routine clinical laboratories has increased significantly. This is mainly due to the ability of MS to provide very specific identification, high sensitivity, and simultaneous analysis of multiple analytes (>100). The coupling of tandem mass spectrometry with gas chromatography (GC) or liquid chromatography (LC) has enabled the rapid expansion of this technology. While applications of MS are used in many clinical areas, therapeutic drug monitoring, drugs of abuse, and clinical toxicology are still the primary focuses of the field. It is not uncommon to see mass spectrometry being used in routine clinical practices for those applications.

**Key words** Clinical laboratory, Mass spectrometry, Liquid chromatography, Gas chromatography, Tandem mass spectrometry, Drugs, Therapeutic drug monitoring, Toxicology, Time-of-flight, Immunoassays

## 1 Introduction

Mass spectrometry, once considered a very specialized and expensive technology for routine use, has made its way in many clinical laboratories in recent years [1, 2]. This rapid growth has been made possible by developments in the technology, the advent of bench top systems, increased ease of operation, reduced capital investment, and more user-friendly software systems.

Therapeutic drug monitoring, testing for drugs of abuse, pain management, and forensic drug testing have been the early adaptors of this technology and still are the main driving force behind the fast growth of the field. Mass spectrometry has been introduced and utilized to overcome the inherent limitations of immunoassays from drug testing due to its high specificity. Enhancements in mass spectrometry continue to improve sensitivity and enable measurement of ever lower concentrations of analytes. GC-MS was the initial MS technique used in clinical laboratories, and introduction

Uttam Garg (ed.), *Clinical Applications of Mass Spectrometry in Drug Analysis: Methods and Protocols*, Methods in Molecular Biology, vol. 1383, DOI 10.1007/978-1-4939-3252-8_1, © Springer Science+Business Media New York 2016

of LC-MS/MS enabled the analysis of many analytes that cannot be easily analyzed by GC and are not suitable for GC-MS. Methods have been reported on a wide array of analytes and the accumulation of experience in the community is available to help interested laboratories overcome the huddles to bring mass spectrometry to their practices to better serve our patients and for the betterment of health care.

The recent developments in mass spectrometry have made it a very attractive platform for clinical practice, yet apprehension due to the complexity of the technology, relatively high capital investment, personnel training, and the requirement for in-house method development and validation present challenges to the implementation of this technology in many routine clinical laboratories.

## 2   Clinical Applications

Immunoassays are commonly used methods for therapeutic drug monitoring and drugs of abuse in the clinical laboratory. Since immunoassays can cause false positive or false negative results due to lack of specificity or cross-reactivity, and immunoassays may not available for a number of drugs, MS has been used for confirmation of immunoassay results [3] and sometime used directly as screening methods. Measurement of small molecule drugs continues to push the development of the technology and to be one of the main driving forces for increasing applications of mass spectrometry in clinical practices. The focus of this volume is for therapeutic drug monitoring and toxicology. Drugs of abuse and therapeutic drugs commonly analyzed by mass spectrometry are listed in Table 1.

Testing for the screening and confirmation of inborn error of metabolism was another early adaptor of mass spectrometry and has played an important role in enhancing the applications of mass spectrometry [4, 5]. Recently, many developments have taken place in new fields of study, particularly endocrinology and hormone testing in clinical labs [6–8].

Although not yet widely found in clinical laboratories, applications of MS are expanding in the analysis of large molecules such as peptides, proteins, lipids, polysaccharides, and DNA [9–11]. Another emerging area is the application of matrix-assisted laser desorption/ionization (MALDI) mass spectrometry to rapid bacterial identification [12–14].

## 3   Fundamentals and Recent Developments of Mass Spectrometry-Based Analysis

A detailed description of mass spectrometry is beyond the scope of this chapter and only a brief description on the fundamentals of the technique is provided. Mass spectrometry is based on the ability to

**Table 1**
**Drugs of abuse and therapeutic drugs commonly assayed by mass spectrometry**

| Drugs of abuse/other | Therapeutic drugs |
|---|---|
| • Amphetamines and related drugs<br>• Barbiturates (amobarbital, butalbital, pentobarbital, phenobarbital, secobarbital, etc.)<br>• Bath salts<br>• Benzodiazepines (alprazolam, diazepam, lorazepam, midazolam, oxazepam, temazepam, clonazepam and their metabolites, etc.)<br>• Buprenorphine<br>• Cocaine and its metabolites<br>• Cannabinoids<br>• Cannabinoids, synthetic<br>• Drug screening, broad spectrum<br>• Ethanol use markers (Ethyl Glucuronide and Ethyl Sulfate)<br>• Ketamine<br>• Methadone and metabolites<br>• Methamphetamine<br>• Nicotine and metabolites<br>• Opiates and opioids (morphine, codeine, hydrocodone, oxycodone, oxymorphone, 6-acetylmorphine, fentanyl, etc.)<br>• Phencyclidine<br>• Propoxyphene<br>• Zolpidem | • Antidepressants (tricyclics and selective serotonin reuptake inhibitors)<br>• Anticoagulants (dabigatran, rivaroxaban, apixaban, warfarin)<br>• Anticonvulsants (lamotrigine, levetiracetam, 10-hydroxycarbazepine, topiramate, zonisamide)<br>• Antipsychotics (haloperidol, fluphenazine, perphenazine, thiothixene)<br>• Busulfan<br>• Cardiac drugs (flecainide, mexiletine, propafenone, amiodarone)<br>• Carisoprodol<br>• 5-Fluorouracil<br>• Ibuprofen<br>• Immunosuppressants (cyclosporine, everolimus, mycophenolic acid, sirolimus, tacrolimus)<br>• Indomethacin<br>• Meprobamate<br>• Methotrexate<br>• Teriflunomide |

influence the motion of charged particles with electric and magnetic fields. This allows separation of charged particles based on their mass-to-charge ($m/z$) ratios. A mass spectrometer can be thought of as an instrument that measures the masses of molecules that have been converted into ions.

**3.1 Mass Spectrometry-Based Analysis Overview**

Analytes need to be either positively or negatively charged (ionized) to be analyzed by a mass spectrometer. In addition, all other charged or potentially ionizable compounds in the sample can potentially be interferences for the analyte of interest. Samples need to be carefully treated before analysis and the sample preparation can be either very simple like "dilute and shoot" or very elaborate. Liquid-liquid or solid-phase extractions are commonly used. While samples can be introduced directly into a mass spectrometer, gas or liquid chromatographic systems for separation are typically used to first isolate the compounds of interest from the matrix.

After separation, the effluent from the chromatograph is ionized. The mass analyzer separates the ions formed based on their mass-to-charge ratio. At the end, the ions are "recorded" in the detector and reported out through the data analysis system.

A schematic diagram of a generic mass spectrometer with different options is shown in Fig. 1.

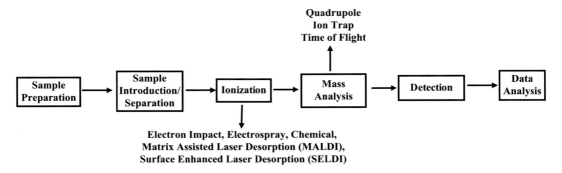

**Fig. 1** Schematic diagram for a mass spectrometry analysis

### 3.2 GC-MS and LC-MS/MS

GC and LC are the two most common chromatography separation techniques coupled to mass spectrometry. GC has been used in clinical laboratories for several decades and LC has gained popularity in recent years mainly due to ease of sample preparation. Using GC-MS or LC-MS/MS, hundreds to thousands of drugs and toxins can be screened in a single analytical run [15]. GC-MS is suitable for analysis of small molecules that are volatile, nonpolar, and thermally stable. Analytes that are heat labile and difficult to derivatize are more suited for LC-MS/MS analysis. While both separation methods typically involves certain types of analyte extraction and possible concentration of the extract, GC often requires lengthy sample derivatization steps for compounds that are not volatile or thermally stable. Simple sample preparation and a broader array of analytes have enabled LC-MS/MS to gain popularity in clinical laboratories. Disadvantages of LC-MS/MS are less reproducible mass spectra, higher maintenance, and high cost.

### 3.3 Types of Mass Spectrometers

A variety of mass analyzers are used in clinical mass spectrometry. The most common types are single quadrupoles, triple quadrupoles, and time-of-flight (TOF) instruments with triple-quadrupole mass spectrometers being the most prevalent in clinical laboratories. Triple quadrupole mass spectrometers offer unique advantages. They are robust and can be used for multiple analytes. They have several scan modes available that are particularly applicable for clinical analyses. The most commonly used scan function is multiple reaction monitoring (MRM).

In an MRM assay, the compound of interest is identified based on the cleavage of a precursor ion to form a fragment ion. One or more fragment ions can be used. Two fragment ions are preferred to increase specificity of the analysis, in which case one fragment ion functions as qualifier ion and the other functions as quantifier ion. With an appropriate internal standard, MRM can generally be used for quantitative analysis.

Time-of-flight instruments bring an additional dimension of analysis due to its high-resolution enabling broad-spectrum screening and confirmation methods. The coupling of MALDI and TOF has enabled the emerging application of microorganism identification in microbiology laboratories.

## 4  Implementing Mass Spectrometry in a Clinical Laboratory

Implementation of mass spectrometry in a clinical laboratory is not an easy undertaking [16–23]. It is a major investment in both financial and human capital for an institution and requires careful planning and diligent execution for a successful outcome. Implementation is a multi-step process and a summary of key considerations is presented in Table 2.

The process typically starts with an assessment of clinical needs, instrument selection, and a financial justification, followed by several other essential tasks such as space planning, site preparation,

**Table 2**
**Major steps in implementing mass spectrometry in a clinical laboratory**

| *Clinical needs* |
| --- |
| • Is the primary consideration |
| • Reduce turnaround time |
| • Control over sample handling process and reduce handling errors |
| *Financial considerations* |
| • Key is to have an institutionally acceptable return on investment (ROI) |
| • Benefits include bringing test in-house and reduce send-out costs |
| • Primary investment is instrument itself |
|    – Capital investment or leasing options |
| • Other investment considerations should include |
|    – Service contract |
|    – Infrastructure and space requirement, and may need renovation |
|    – Cost for interfacing to the LIS if desirable |
|    – Ongoing operating cost (e.g., high-grade reagents, special reagents, gas) |
| *Instrument selection* |
| • Based on intended analyses and economics |
| • Site visit and communication with colleagues and vendors |
| • Service availability and response time for service requests |
| *Assay selection* |
| • Based on type of instrumentation, analytes, and clinical needs |
| • Literature search and communication with colleagues |
| • Consider lab staff experience and training |

(continued)

**Table 2**
**(continued)**

| *Infrastructure planning* |
| --- |
| • Space for instrumentation and HPLC<br>• Gas supplies: compressor air, nitrogen gas dewars, or nitrogen generator<br>• Ventilation and noise blocking<br>• Lab space rearrangements (e.g., fixed vs. movable bench)<br>• Dedicated electric system and uninterrupted power supply<br>• IT support and data backup |
| *Staff and personnel training* |
| • Essential for a successful implementation<br>• Is an ongoing process<br>• Onsite training with manufacturers<br>• Online training courses<br>• Conferences workshops, symposia, and short courses |
| *Method development and validation* |
| • Meets CLIA requirements for high complex testing<br>• Use highest grade reagents available (MS grade or at least HPLC grade)<br>• Choose proper internal standards<br>• Validation shall include<br>  – Precision<br>  – Accuracy<br>  – Analytical sensitivity<br>  – Analytical range and reportable range<br>  – Reference interval validation<br>  – Stability<br>  – Specificity and interference testing |

installation, personnel training, and method evaluation. In total, the whole process can take from 6–9 months to 1 year. After that, in-house method development and validation bring another set of challenges. Mass spectrometry assays are considered high-complexity under CLIA and fall under the category of laboratory-developed tests. This section sheds some light onto these steps with the hope of saving the audience some headaches in the implementation phase.

**4.1 Assay and Instrument Selection**

A large body of evidence has indicated the value of mass spectrometry for clinical practice and many well-established tests are available for labs to consider [24–26]. For a lab new to mass spectrometry, clinical needs justification becomes a question of which tests to bring in-house and which methods to choose. Consultation with in-house physicians and advocates will ease decision making and increase acceptance.

While modern mass spec companies all provide high quality products, each vendor has its own unique "temperament." Consulting with colleagues and site visits can help narrow down the choices of vendors for further investigation. In addition, service availability and response time play an important role in decision making and should be considered very seriously.

**4.2 Financial Justification**

Financial justification is one of the initial challenges of bringing mass spectrometry technology to a laboratory. The financial justification can be approximated as a return on investment (ROI) estimation. The cost of instruments (ranges from low US$200,000 to US$4–500,000) and the savings from bringing tests in house on mass spectrometry are obvious considerations. Several other factors are not necessarily obvious for people who are new in this field and are important for the total financial estimation. These factors include the cost of an annual service contract, cost for infrastructure planning and space renovation, and operating costs. Each one of these cost can be executed with different options, which can make a large difference in the final financial commitment. While the typical practice has been to acquire instrumentation through the standard equipment capitalization process by purchasing the instrument upfront, reagent leasing and equipment leasing are other options. It is worth the time to go through the options for any given test and projected volume.

It is recommended to have a full-service contract for most laboratories to cover regular maintenance and repairs. Manufacturers often provide discounts for pre-purchased service contracts after the warranty period or a discount for purchasing multiple-year contracts.

Mass spectrometers require specific infrastructure elements, such as particular voltage/current requirements, high purity gas, ventilation and noise blocking system which are not necessarily in place at most institutions. Communication with the vendors is highly recommended to ensure that the infrastructure and space planning meet the vendor's specific requirements.

Besides the initial fixed cost, running an assay is relatively inexpensive. Nonetheless, several factors such as consumables for sample preparation, reagents, internal standards, quality controls, gas supplies, data analysis, and data reporting should be considered. MS-grade reagents can be expensive, and it is important to get the highest grade of reagents for MS assays. If they are not available, HPLC-grade reagents can sometimes be substituted. Internal standards are unique to mass spectrometry assays and can be very expensive.

The majority of mass spectrometry data still includes some manual processing, which is labor intensive. Many manufacturers are able and willing to assist in interfacing the data system to the laboratory information system (LIS), which can reduce the reporting time significantly. If this is desirable, the cost for interfacing is another consideration.

Reimbursement rates for mass spectrometry assays are reasonable and the variable operating cost is relatively low. Despite the costs, the addition of mass spectrometric analyses to a laboratory in most cases can provide a reasonable ROI and can be financially justified.

### 4.3 Staff Training and Human Resources

Having properly trained staff for routine analysis and a skilled analyst for troubleshooting and method development are crucial to successful implementation. Hiring staff with proper training for mass spectrometry has been one of the major challenges to implementation of this technology in clinical laboratories. However, with dedicated instrument time and an experienced trainer, a good medical technologist can be trained to perform mass spectrometric analyses within a few weeks. When method development is involved, longer and more sophisticated training is required.

Training can be obtained in several ways. Instrument manufacturers can provide training during the installation, through their onsite live training classes or online tutorials. Many online training materials and tutorials are available from other organizations such as the American Association for Clinical Chemistry, and colleagues are always good resources. There is nothing, however, that can substitute for the experience gained from sitting in front of the instrument.

### 4.4 Method Development and Validation

Mass spectrometry assays are laboratory-developed tests, which require sophisticated validation processes. Most assays are developed in-house and standardization of the process is very important. The recent CLSI guideline is a good reference [27]. In addition to the typical validation steps for clinical assays including limit of detection, limit of quantitation, accuracy, analytical measurement range and clinical reportable range, reagent and analyte stability, reference range determination, and interference, mass spectrometric analyses require additional validation steps. A particular requirement of mass spectrometric analyses is an assessment of ion suppression. There are different ways to do ion suppression assessment, and compromise between the amount of sample cleanup and the level of ion suppression tolerable is often necessary. In many cases, the use of an internal standard can help correct for moderate levels of ion suppression.

The internal standard plays an important role in providing accurate results with mass spectrometry. The internal standard should be chemically similar to the analyte but with a different molecular weight. The heavy isotope labeled compounds with C13 and N15 have become the typical choice. A mass difference between the analyte of interest and the internal standard of at least 3 mass units is desirable, although a difference of at least 5 is preferred to completely reduce cross talk.

## 5 Conclusion

In recent years mass spectrometry has emerged as an important tool in laboratory medicine. While it has been applied in many clinical areas, therapeutic drug monitoring and toxicology remain as key applications and continue to be the driving force to push through challenges to increase the applications of mass spectrometry in clinical practices. Many methods have published in recent years in these areas and for analytes that were once considered unlikely for MS. Mass spectrometry remains a new, exciting, and sophisticated technology. Its implementation requires careful planning and diligent execution to ensure success in routine practices. This volume has collected many methods currently used in clinical laboratories with a sufficient level of detail to allow straightforward implementation. We hope that this will facilitate the adaption of this technology in clinical practices.

### References

1. Hammett-Stabler CA, Garg U (2010) The evolution of mass spectrometry in the clinical laboratory. Methods Mol Biol 603:1–7
2. Strathmann FG, Hoofnagle AN (2011) Current and future applications of mass spectrometry to the clinical laboratory. Am J Clin Pathol 136:609–616
3. Maurer HH (2007) Current role of liquid chromatography-mass spectrometry in clinical and forensic toxicology. Anal Bioanal Chem 388:1315–1325
4. Garg U, Dasouki M (2006) Expanded newborn screening of inherited metabolic disorders by tandem mass spectrometry: clinical and laboratory aspects. Clin Biochem 39:315–332
5. Jones PM, Bennett MJ (2002) The changing face of newborn screening: diagnosis of inborn errors of metabolism by tandem mass spectrometry. Clin Chim Acta 324:121–128
6. Pagotto U, Fanelli F, Pasquali R (2013) Insights into tandem mass spectrometry for the laboratory endocrinology. Rev Endocr Metab Disord 14:141
7. Vogeser M, Parhofer KG (2007) Liquid chromatography tandem-mass spectrometry (LC-MS/MS)—technique and applications in endocrinology. Exp Clin Endocrinol Diabetes 115:559–570
8. Soldin SJ, Soldin OP (2009) Steroid hormone analysis by tandem mass spectrometry. Clin Chem 55:1061–1066
9. Jimenez CR, Verheul HM (2014) Mass spectrometry-based proteomics: from cancer biology to protein biomarkers, drug targets, and clinical applications. Am Soc Clin Oncol Educ Book e504–e510
10. Li Y, Song X, Zhao X, Zou L, Xu G (2014) Serum metabolic profiling study of lung cancer using ultra high performance liquid chromatography/quadrupole time-of-flight mass spectrometry. J Chromatogr B Analyt Technol Biomed Life Sci 966:147–153
11. Whiteaker JR (2010) The increasing role of mass spectrometry in quantitative clinical proteomics. Clin Chem 56:1373–1374
12. Ho YP, Reddy PM (2011) Advances in mass spectrometry for the identification of pathogens. Mass Spectrom Rev 30:1203–1224
13. Lagace-Wiens P (2015) Matrix-assisted laser desorption/ionization time of flight mass spectrometry (MALDI-TOF/MS)-based identification of pathogens from positive blood culture bottles. Methods Mol Biol 1237:47–55
14. Luan J, Yuan J, Li X, Jin S, Yu L, Liao M, Zhang H, Xu C, He Q, Wen B et al (2009) Multiplex detection of 60 hepatitis B virus variants by maldi-tof mass spectrometry. Clin Chem 55:1503–1509
15. Ojanpera I, Kolmonen M, Pelander A (2012) Current use of high-resolution mass spectrometry in drug screening relevant to clinical and forensic toxicology and doping control. Anal Bioanal Chem 403:1203–1220
16. Clarke W, Rhea JM, Molinaro R (2013) Challenges in implementing clinical liquid chromatography-tandem mass spectrometry methods—the light at the end of the tunnel. J Mass Spectrom 48:755–767

17. Roux A, Lison D, Junot C, Heilier JF (2011) Applications of liquid chromatography coupled to mass spectrometry-based metabolomics in clinical chemistry and toxicology: a review. Clin Biochem 44:119–135

18. Wu AH, French D (2013) Implementation of liquid chromatography/mass spectrometry into the clinical laboratory. Clin Chim Acta 420:4–10

19. Armbruster DA, Overcash DR, Reyes J (2014) Clinical Chemistry Laboratory Automation in the 21st century—Amat Victoria curam (Victory loves careful preparation). Clin Biochem Rev 35:143–153

20. Vogeser M, Kirchhoff F (2011) Progress in automation of LC-MS in laboratory medicine. Clin Biochem 44:4–13

21. Vogeser M, Seger C (2010) Pitfalls associated with the use of liquid chromatography-tandem mass spectrometry in the clinical laboratory. Clin Chem 56:1234–1244

22. Carvalho VM (2012) The coming of age of liquid chromatography coupled to tandem mass spectrometry in the endocrinology laboratory. J Chromatogr B Analyt Technol Biomed Life Sci 883–884:50–58

23. Himmelsbach M (2012) 10 Years of MS instrumental developments—impact on LC-MS/MS in clinical chemistry. J Chromatogr B Analyt Technol Biomed Life Sci 883–884:3–17

24. Kortz L, Dorow J, Ceglarek U (2014) Liquid chromatography-tandem mass spectrometry for the analysis of eicosanoids and related lipids in human biological matrices: a review. J Chromatogr B Analyt Technol Biomed Life Sci 964:1–11

25. van den Ouweland JM, Vogeser M, Bacher S (2013) Vitamin D and metabolites measurement by tandem mass spectrometry. Rev Endocr Metab Disord 14:159–184

26. Peters FT, Remane D (2012) Aspects of matrix effects in applications of liquid chromatography-mass spectrometry to forensic and clinical toxicology—a review. Anal Bioanal Chem 403:2155–2172

27. CLSI (2014). In: Clinical and Laboratory Standards Institute (ed.), CLSI Document C62-A, vol 62-A. CLSI, Wayne, PA

# Chapter 2

# Quantitation of Flecainide, Mexiletine, Propafenone, and Amiodarone in Serum or Plasma Using Liquid Chromatography-Tandem Mass Spectrometry (LC-MS/MS)

## Matthew H. Slawson and Kamisha L. Johnson-Davis

### Abstract

Flecainide, mexiletine, propafenone, and amiodarone are antiarrhythmic drugs that are used primarily in the treatment of cardiac arrhythmias. The monitoring of the use of these drugs has applications in therapeutic drug monitoring and overdose situations. LC-MS/MS is used to analyze plasma/serum extracts with loxapine as the internal standard to ensure accurate quantitation and control for any potential matrix effects. Positive ion electrospray is used to introduce the analytes into the mass spectrometer. Selected reaction monitoring of two product ions for each analyte allows for the calculation of ion ratios which ensures correct identification of each analyte, while a matrix matched calibration curve is used for quantitation.

**Key words** Flecainide, Mexiletine, Propafenone, Amiodarone, Plasma, Serum, UPLC, Mass spectrometry

## 1 Introduction

Cardiac arrhythmias are caused by abnormal activity of the endogenous pacemaker or by abnormalities associated with impulse propagation. Several drugs exist to pharmacologically treat cardiac arrhythmia in lieu or in addition to more invasive procedures such as artificial pacemakers, ablations, or other surgical interventions. Mexilitene (e.g., Mexitil) is a Class IB sodium channel blocker related to lidocaine. Flecainide (e.g., Tambocor) and propafenone (e.g., Rhythmol) are potent sodium channel blocker of Class IC. In addition, amiodarone (e.g., Cordarone) exhibits properties of all four classes of antiarrhythmic drugs and also possess anti-anginal properties [1, 2].

This method describes an analytical method to measure these four antiarrhythmic drugs in human serum/plasma by precipitating serum/plasma proteins and collecting the supernatant for analysis.

Uttam Garg (ed.), *Clinical Applications of Mass Spectrometry in Drug Analysis: Methods and Protocols*, Methods in Molecular Biology, vol. 1383, DOI 10.1007/978-1-4939-3252-8_2, © Springer Science+Business Media New York 2016

The supernatant is injected onto the LC-MS/MS. Qualitative identification is made using unique MS/MS transitions, ion ratios of those transitions and chromatographic retention time. Quantitation is performed using a daily calibration curve of prepared calibration samples and using peak area ratios of analyte to internal standard to establish the calibration model. Patient sample concentrations are calculated based on the calibration model's mathematical equation. Quantitative accuracy is monitored with QC samples independently prepared with known concentrations of analyte and comparing the calculated concentration with the expected concentration [3–6].

## 2 Materials

### 2.1 Samples

1. Pre-dose (trough) draw—At steady-state concentration for serum/plasma. Separate serum or plasma from cells within 2 h of collection.

2. Collect in plain Red tube (Lavender or Pink top tubes also acceptable). Avoid gel or other separator tubes.

3. Specimens can be stored at ambient temperature for 4 h (propafenone), 48 h (mexilitine), 4 weeks (amiodarone), 6 weeks (flecainide); refrigerated for 5 days (mexilitiene) or at least 4 weeks (propafenone, flecainide, amiodarone); and at least 4 weeks frozen prior to analysis.

### 2.2 Reagents

1. Clinical Laboratory Reagent Water (CLRW).

2. Mobile Phase A (CLRW with 0.1 % formic acid): 1.0 mL of concentrated formic acid in CLRW q.s. to 1.0 L in volumetric flask.

3. Mobile Phase B: Acetonitrile with 0.1 % formic acid: 1.0 mL of concentrated formic acid in LC-MS-grade acetonitrile q.s. to 1.0 L in volumetric flask.

### 2.3 Standards and Calibrators

1. Flecainide, 1.0 mg/mL stock standard prepared in Methanol (Cerilliant, Round Rock, TX).

2. Propafenone 1.0 mg/mL stock standard prepared in methanol from commercially available powder (Sigma-Aldrich, St. Louis, MO).

3. Mexilitene 1.0 mg/mL stock standard prepared in methanol from commercially available powder (Sigma-Aldrich, St. Louis, MO).

4. Amiodarone 1.0 mg/mL stock standard prepared in methanol (Cerilliant, Round Rock, TX).

5. Prepare working calibrators to prepare 25 mL of each using volumetric glassware. Add approximately 10 mL certified negative

**Table 1**
**Preparation of calibrators**

| Calibrator | Volume of each 1 mg/mL solution (µL) | Final [] (µg/mL) |
|---|---|---|
| 1 | 2.5 | 0.1 |
| 2 | 7.5 | 0.3 |
| 3 | 15 | 0.6 |
| 4 | 25 | 1 |
| 5 | 75 | 3 |
| 6 | 150 | 6 |

The total volume is made to 25 mL with drug-free human serum/plasma

plasma/serum to a labeled volumetric flask. Add the appropriate volume as shown in Table 1 of each solution described in #1–5 above to the flask; q.s. to 25 mL using certified negative serum/plasma. Add a stir bar and stopper and mix for at least 30 min at room temperature. Aliquot as appropriate for future use. Store aliquots frozen, stable for 1 year. This volume can be scaled up or down as appropriate (*see* **Notes 1** and **2**).

**2.4 Controls and Internal Standard**

1. Controls: May be purchased from a third party and prepared according to the manufacturer. They can also prepared in-house independently from calibrators' source material using Table 1 as a guideline (*see* **Note 1**).

2. Internal standard/protein precipitation solution:

    (a) Prepare a 0.5 mg/mL stock solution of internal standard by adding 2.5 mg loxapine succinate salt (Sigma-Aldrich, St. Louis, MO) to a 5 mL volumetric flask, q.s. to volume with methanol, add a stir bar and stopper, and stir at room temperature for at least 30 min until equilibrated.

    (b) Add 0.25 L of acetonitrile to a 500 mL volumetric flask, add 2.5 mL of the stock internal standard solution made in **item 2a** (above) and q.s the flask with methanol to 500 mL. Add a stir bar and stopper and allow mixing at room temperature for at least 30 min until equilibrated. Aliquot as needed for use in this assay (volumes can be scaled up or down as appropriate). Store frozen, stable for 1 year (*see* **Notes 1** and **2**).

**2.5 Supplies and Equipment**

1. Transfer/aliquoting pipettes and tips.

2. 1.5 mL polypropylene microcentrifuge tubes with caps.

3. Instrument compatible autosampler vials with injector appropriate caps.

4. Acquity HSS T3 1.8 μm, 2.1 × 50 mm UPLC column (Waters, Milford, MA).

5. Multi-tube Vortex mixer (e.g., VWR VX-2500).

6. Foam rack(s) compatible with both microcentrifuge tubes and multi-tube vortex mixer.

7. Centrifuge capable of 18,000 × *g* that will accommodate micro-centrifuge tubes.

8. Waters Acquity TQD UPLC-MS/MS system (Milford, MA).

## 3  Methods

### 3.1  Stepwise Procedure

1. Briefly vortex or invert each sample to mix.

2. Aliquot 50 μL of each patient sample, calibrator and QC into appropriately labeled micrcentrifuge tubes.

3. Add 500 μL of internal standard/precipitation solution to each vial.

4. Cap each tube and vortex vigorously for 30 s.

5. Centrifuge for ~10 min at ~18,000 × *g* (*see* **Note 3**).

6. Transfer the contents of each tube (from **steps 2** to **5**) to an autosampler vial and cap.

7. Analyze on LC-MS/MS.

### 3.2  Instrument Operating Conditions

1. Table 2 summarizes typical LC conditions.

2. Table 3 summarizes typical MS conditions.

3. Table 4 summarizes typical MRM conditions.

Each instrument should be individually optimized for best method performance.

### 3.3  Data Analysis

1. Representative MRM chromatograms of each antiarrhythmic and internal standard in plasma are shown in Fig. 1a–e.

2. The dynamic range for this assay is 0.1–6 μg/mL. Samples exceeding this range can be diluted 5× or 10× as needed to achieve an accurate calculated concentration, if needed.

3. Data analysis is performed using the QuanLynx or TargetLynx software to integrate peaks, calculate peak area ratios, and construct calibration curves using a linear $1/x$ weighted fit ignoring the origin as a data point. Sample concentrations are then calculated using the derived calibration curves (*see* **Note 4**).

4. Calibration curves should have an $r^2$ value $\geq 0.99$.

5. Typical imprecision is <15 % both inter- and intra-assay.

**Table 2**
**Typical HPLC conditions**

| | |
|---|---|
| Weak wash | Mobile phase A |
| Strong wash | Mobile phase B |
| Seal wash | Mobile phase A |
| Injection volume | 2 µL |
| Vacuum degassing | On |
| Temperature | 30 °C |
| A Reservoir | 0.1 % HCOOH in CLRW |
| B Reservoir | 0.1 % HCOOH in acetonitrile |
| Gradient table | |

| Step | Time (min) | Flow (µL/min) | A (%) | B (%) | Curve[a] |
|---|---|---|---|---|---|
| 0 | 0 | 650 | 70 | 30 | 1 |
| 1 | 1 | 650 | 55 | 45 | 6 |
| 2 | 1.3 | 650 | 10 | 90 | 6 |
| 3 | 1.55 | 650 | 70 | 30 | 11 |

[a]Nonlinear gradient curves common to Waters systems

**Table 3**
**Typical mass spectrometer conditions**

| Parameter | Value |
|---|---|
| Capillary (kV) | 0.8 |
| Cone (V) | 20–40 |
| Extractor (V) | 3 |
| RF (V) | 0.1 |
| Desolvation temp | 450 |
| Desolvation gas | 900 |
| Cone gas | 30 |
| Collision gas | 0.25 |
| Scan mode | MSMS |
| Polarity | Positive |
| Ion source | ESI |
| Resolution Q1 | Unit |
| Resolution Q3 | Unit |
| Dwell (s) | 0.045 |

**Table 4**
**Typical MRM conditions**

| Analyte | Precursor | Product (quant) | Product (qual.) |
|---------|-----------|-----------------|-----------------|
| Flecainide | 415.1 | 398.1 | 301.1 |
|  | Collision energy | 25 | 35 |
| Mexiletine | 180.1 | 58.1 | 105.1 |
|  | Collision energy | 10 | 20 |
| Propafenone | 342.1 | 98.1 | 72.1 |
|  | Collision energy | 25 | 30 |
| Amiodarone | 646.1 | 58.1 | 72.1 |
|  | Collision energy | 45 | 30 |
| Loxapine (internal standard) | 328.1 | 84.1 | 297.1 |
|  | Collision energy | 25 | 25 |

6. An analytical batch is considered acceptable if chromatography is acceptable and QC samples calculate to within 20 % if their target values and ion ratios are within 20 % of the calibration curve ion ratios.

## 4   Notes

1. Validate/verify all calibrators, QCs, internal standard, and negative matrix pools before placing into use.

2. Carefully add methanolic solutions to plasma to avoid excessive protein precipitation. If desired, methanol can be evaporated in the volumetric flask prior to addition of plasma/serum.

3. Time and speed of centrifugation step can be optimized to ensure that a firm pellet is formed so as not to transfer any precipitate to autosampler vial.

4. Loxapine shows good recovery and a retention time intermediate to the other analytes making it a good compromise internal standard for all four antiarrythmics. Ion suppression studies (data not shown) indicate that this I.S. offers control of matrix effects under the conditions described. Structural and/or isotopically labeled analogs of each drug may be utilized if desired.

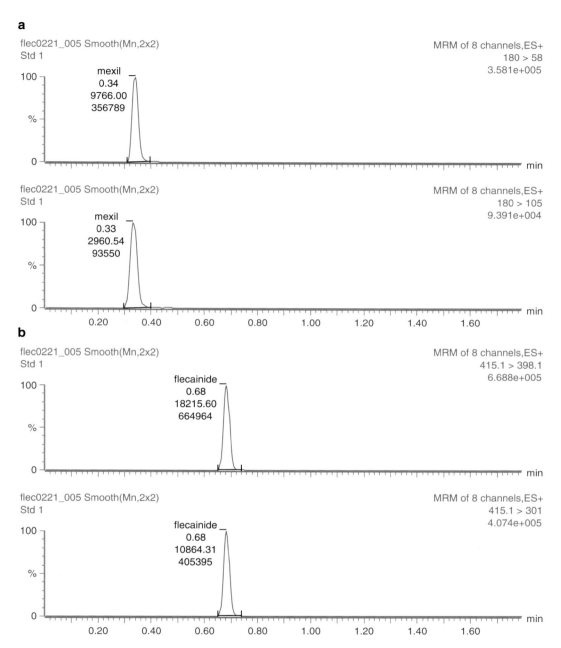

**Fig. 1** Typical MRM chromatograms for (**a**) mexilitene (0.33 min), (**b**) flecainide (0.68 min), (**c**) loxapine (0.71 min), (**d**) propafenone (0.86 min), (**e**) amiodarone (1.46 min) in plasma. 0.3 ng/mL (see text) extracted from fortified human plasma and analyzed according to the described method

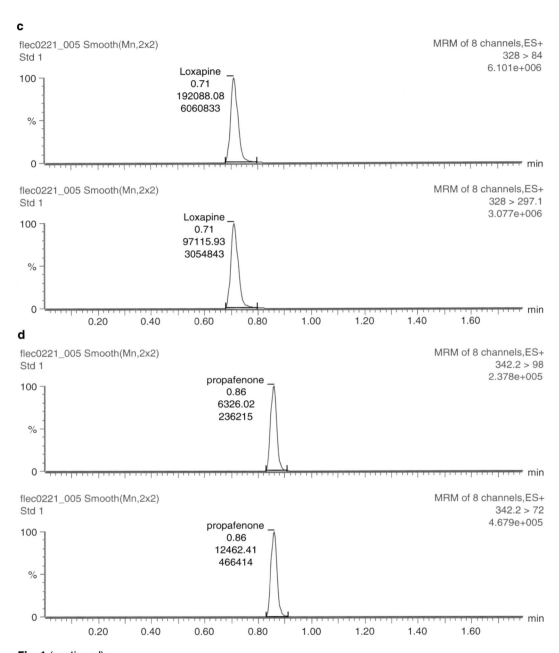

**Fig. 1** (continued)

**e**

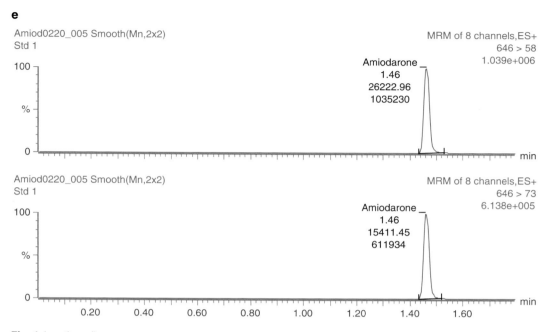

**Fig. 1** (continued)

## References

1. Goodman LS, Brunton LL, Chabner B, Knollmann BC (2011) Goodman & Gilman's pharmacological basis of therapeutics, 12th edn. McGraw-Hill, New York

2. Baselt RC (2011) Disposition of toxic drugs and chemicals in man, 9th edn. Biomedical Publications, Seal Beach, CA

3. Hofmann U, Pecia M, Heinkele G, Dilger K, Kroemer HK, Eichelbaum M (2000) Determination of propafenone and its phase I and phase II metabolites in plasma and urine by high-performance liquid chromatography-electrospray ionization mass spectrometry. J Chromatogr B Biomed Sci Appl 748(1):113–123

4. Yan H, Xiang P, Bo J, Shen M (2007) Determination of mexiletine in human blood by liquid chromatography-tandem mass spectrometry. Fa yi xue za zhi 23(6):441–443

5. Kollroser M, Schober C (2002) Determination of amiodarone and desethylamiodarone in human plasma by high-performance liquid chromatography-electrospray ionization tandem mass spectrometry with an ion trap detector. J Chromatogr B Analyt Technol Biomed Life Sci 766(2):219–226

6. Mano Y, Asakawa Y, Kita K, Ishii T, Hotta K, Kusano K (2015) Validation of an ultra-performance liquid chromatography-tandem mass spectrometry method for the determination of flecainide in human plasma and its clinical application. Biomed Chromatogr. doi:10.1002/bmc.3437

# Chapter 3

# Quantitation of the Oral Anticoagulants Dabigatran, Rivaroxaban, Apixaban, and Warfarin in Plasma Using Ultra-Performance Liquid Chromatography with Tandem Mass Spectrometry (UPLC-MS/MS)

## Jaime H. Noguez and James C. Ritchie

## Abstract

This chapter describes a method to measure the oral anticoagulants dabigatran, rivaroxaban, apixaban, and warfarin in plasma samples using ultra-performance liquid chromatography combined with tandem mass spectrometry (UPLC-MS/MS). The instrument is operated in multiple reaction monitoring (MRM) mode with an electrospray ionization (ESI) source in positive ionization mode. Samples are extracted with a 90:10 methanol/0.1 N hydrochloric acid solution containing stable isotope-labeled internal standards for each analyte. After centrifugation the supernatant is transferred to a mass spectrometry vial, injected onto the UPLC-ESI-MS/MS, and quantified using an eight-point calibration curve.

**Key words** Anticoagulant, Dabigatran, Rivaroxaban, Apixaban, Warfarin, Mass spectrometry

## 1  Introduction

Millions of patients worldwide are prescribed oral anticoagulant therapy. Until recently vitamin K antagonists, such as warfarin (the most commonly prescribed drug in this class), were the only oral anticoagulant drugs available to clinicians. The US Food and Drug Administration recently approved dabigatran etexilate, rivaroxaban, and apixaban for the prevention and treatment of thrombosis [1]. These drugs represent a new class of oral anticoagulants referred to as *n*on-vitamin K *o*ral *a*nti*c*oagulants (NOACs) and unlike the traditionally used therapies they target and selectively inhibit specific enzymes in the coagulation cascade. Dabigatran etexilate, a prodrug for dabigatran, is a direct thrombin inhibitor; while rivaroxaban and apixaban directly inhibit factor Xa [2–4]. Their direct mechanisms of action provide a number of benefits such as rapid onset of activity, short half-life, and a wide therapeutic range [2–4].

Uttam Garg (ed.), *Clinical Applications of Mass Spectrometry in Drug Analysis: Methods and Protocols*, Methods in Molecular Biology, vol. 1383, DOI 10.1007/978-1-4939-3252-8_3, © Springer Science+Business Media New York 2016

Although the NOACs were designed for use without the need for routine laboratory monitoring, due to their predictable pharmacological profiles, there are a number of situations when identification and quantitation may be required for clinical decision making, such as: determining whether drug failure is the cause of an adverse event (thrombotic/hemorrhagic), checking compliance with the prescribed therapy, perioperative patient management, suspected overdose, identifying drugs in unconscious or incoherent patients, checking patients with acute ischemic stroke prior to administration of thrombolytic therapy, and monitoring patients at risk for drug accumulation. The plasma concentrations of these new drugs and warfarin can be measured simultaneously with good accuracy and precision using UPLC-ESI-MS/MS in positive ionization mode. Simultaneous measurement of these four drugs may be especially useful for monitoring patients transitioning between oral anticoagulant therapies.

## 2   Materials

### 2.1   Samples

Acceptable sample types are heparin, citrate, or EDTA plasma. Samples are stable for 4 days when refrigerated and up to 12 months when frozen at –80 °C.

### 2.2   Solvents and Reagents

1. Mobile phase A (0.1 % formic acid in water): Add 1 mL of formic acid to a 1 L volumetric flask, bring to volume with water, and mix. Stable at room temperature (18–24 °C) for 1 year.

2. Mobile phase B (0.1 % formic acid in methanol): Add 1 mL of formic acid to a 1 L volumetric flask, bring to volume with methanol, and mix. Stable at room temperature (18–24 °C) for 1 year.

3. Human drug-free pooled plasma.

### 2.3   Standards and Calibrators

1. Primary standard solutions (1 mg/mL in 9:1 methanol/DMSO): Dabigatran and Apixaban (Alsachim, Strasbourg, France), Rivaroxaban (SynFine Research, Inc., Ontario, Canada), (+/−) Warfarin (Sigma Aldrich, St. Louis, MO).

2. Combined working standard (10 μg/mL in methanol): Transfer 100 μL of each primary standard (dabigatran, apixaban, rivaroxaban, (+/−) warfarin) to a 10 mL volumetric flask and fill with methanol. The combined working standard is stable for 1 year at –80 °C.

3. Calibrators are prepared according to Table 1 using 10 mL volumetric flasks. The calibrators are stable for 1 year at –80 °C.

**Table 1**
**Preparation of calibrators**

| Calibrator | µL of combined working standard (10 µg/mL) | Drug-free plasma (mL) | Concentration (ng/mL) |
|---|---|---|---|
| 1 | 0 | 10.000 | 0 |
| 2 | 5 | 9.995 | 5 |
| 3 | 25 | 9.975 | 25 |
| 4 | 50 | 9.950 | 50 |
| 5 | 125 | 9.875 | 125 |
| 6 | 250 | 9.750 | 250 |
| 7 | 500 | 9.500 | 500 |
| 8 | 1000 | 9.000 | 1000 |

The total volume is made to 10 mL with drug-free human plasma
Calibrators are stable for 1 year at −80 °C

*2.4 Internal Standards and Quality Controls*

1. Primary internal standard (IS) solutions (1 mg/mL in 9:1 methanol/DMSO): [$^{13}C_6$] Dabigatran, [$^{13}C$, $^2H_7$]-Apixaban, and [$^{13}C_6$]-Rivaroxaban (Alsachim, Strasbourg, France), (+/−) Warfarin-d5 (C/D/N Isotopes, Inc., Quebec, Canada).

2. Secondary IS solutions, 100 µg/mL in methanol: Transfer 0.5 mL of each primary standard to a separate 5 mL volumetric flask and fill to volume with methanol. The secondary IS solutions are stable for 1 year at −80 °C.

3. Working IS solution, (400 ng/mL [$^{13}C_6$]-Rivaroxaban, 17 ng/mL [$^{13}C_6$]-Dabigatran, 2200 ng/mL [$^{13}C$, $^2H_7$]-Apixaban, and 400 ng/mL (+/−) Warfarin-d5 in 90:10 Methanol: 0.1 N HCl): Transfer 400 µL of [$^{13}C_6$]-Rivaroxaban secondary IS solution, 17 µL [$^{13}C_6$]-Dabigatran secondary IS solution, 2.2 mL [$^{13}C$, $^2H_7$]-Apixaban secondary IS solution, and 400 µL (+/−) Warfarin-d5 secondary IS solution into a single 100 mL volumetric flask. Add 10 mL of 0.1 N HCl then fill to volume with methanol. The working IS solution is stable for 6 months refrigerated.

4. Prepare low (50 ng/mL), medium (350 ng/mL) and high (650 ng/mL) controls according to Table 2 by spiking drug-free human plasma with a different set of dabigatran, rivaroxaban, apixaban, and warfarin stock solutions. Transfer aliquots to microcentrifuge tubes for storage at −80 °C. The controls are stable for 1 year.

**Table 2**
**Preparation of in-house plasma controls**

| Control | µL of combined working standard (10 µg/mL) | Drug-free plasma (mL) | Concentration (ng/mL) |
|---|---|---|---|
| Low | 50 | 9.95 | 50 |
| Medium | 350 | 9.65 | 350 |
| High | 650 | 9.35 | 650 |

Calibrators are stable for 1 year at –80 °C

### 2.5 Supplies and Equipment

1. ZORBAX SB-CN column, 3.0 mm × 50 mm, 1.8 µm particle size, 600 bar (Agilent Technologies, Santa Clara, CA).
2. Waters ACQUITY UPLC System with autosampler (Waters Corp., Milford, MA).
3. Waters Xevo TQ MS equipped with MassLynx analytical software (Waters Corp., Milford, MA).

## 3 Methods

### 3.1 Stepwise Procedure

1. Pipet 100 µL of each standard, control, and sample to the appropriately labeled microcentrifuge tube.
2. Add 250 µL of working internal standard solution to each tube, mix well on vortex mixer, and then let sit for 10 min (*see* **Note 1**).
3. Centrifuge for 7 min at 14,000 × *g* (*see* **Note 2**).
4. Transfer supernatant to an appropriately labeled autosampler screw-topped MS sample vial (*see* **Note 3**).
5. Place vial into HPLC autosampler.

### 3.2 Instrument Operating Conditions

1. Inject 10 µL on the LC-MS/MS system for analysis.
2. The instrument's operating conditions are given in Table 3.

### 3.3 Data Analysis

1. A representative UPLC-MS/MS chromatogram is shown in Fig. 1.
2. The data are analyzed using MassLynx Software (Waters Corp., Milford, MA). Standard curves are generated based on linear aggression of the analyte/IS peak ratio versus analyte concentration using the quantifying ions listed in Table 4.
3. A typical calibration curve has a correlation coefficient ($R^2$) of >0.99.

**Table 3**
**UPLC-ESI-MS/MS operating conditions**

| A. UPLC[a] | | |
|---|---|---|
| Column temp. (°C) | 35 | |
| Flow (mL/min) | 0.4 | |
| Gradient | Time (min) | Mobile phase A (%) |
| | 0 | 100 |
| | 2.00 | 50 |
| | 4.00 | 34 |
| | 5.00 | 0 |
| | 5.50 | 100 |
| B. MS/MS tune settings[b] | | |
| Capillary (kV) | 0.5 | |
| Source temp. (°C) | 150 °C | |
| Desolvation temp. (°C) | 500 °C | |
| Cone gas flow (L/h) | 1 | |
| Desolvation gas flow (L/h) | 1000 | |
| LM 1 resolution | 2.56 | |
| LM 2 resolution | 2.80 | |
| HM 1 resolution | 14.68 | |
| HM 2 resolution | 14.86 | |

[a]Mobile phase A, 0.1 % formic acid in water; mobile phase B, 0.1 % formic acid in methanol
[b]Tune settings may vary slightly between instruments

4. An analytical run is considered acceptable if the calculated control concentrations are within two standard deviations of the target values and the ratios of qualifier ions to quantifying ions are within 10 % of the ion ratios for the calibrators.

5. Quantitation is linear to 1000 ng/mL and lower limits of quantitation are as follows: dabigatran 5.0 ng/mL, warfarin 2.5 ng/mL, apixaban 1.5 ng/mL, and rivaroxaban 0.5 ng/mL. Samples in which the drug concentrations exceed the upper limit of quantitation should be diluted with drug-free plasma and retested.

6. The intra- and inter-assay variation (%CV) is <10 % for all analytes over the entire range.

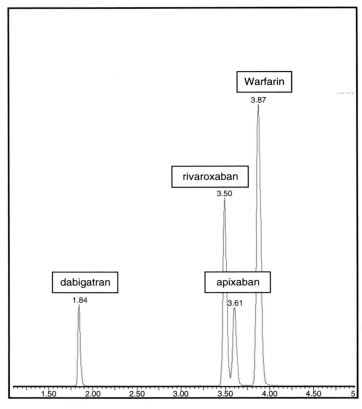

**Fig. 1** UPLC-ESI-MS/MS chromatogram of the oral anticoagulants dabigatran, rivaroxaban, apixaban, and warfarin

**Table 4**
**Quantifying and qualifying ions**

| Compound | Precursor ion (M + H) | Product ions[a] (quantitative, qualifier) | Cone (V) | Collision (eV) | Dwell (s) |
|---|---|---|---|---|---|
| Warfarin | 309.00 | 163.00, 251.05 | 20.00 | 15.00 | 0.025 |
| Warfarin-d5 | 314.2 | 163.00, 251.20 | 20.00 | 17.00 | 0.025 |
| Rivaroxaban | 436.20 | 144.90, 231.20 | 35.00 | 25.00 | 0.025 |
| [$^{13}C_6$]-Rivaroxaban | 442.00 | 144.90 | 35.00 | 30.00 | 0.025 |
| Apixaban | 460.34 | 199.05, 76.97 | 40.00 | 39.00 | 0.025 |
| [$^{13}C, ^2H_7$]-Apixaban | 482.20 | 199.05 | 50.00 | 40.00 | 0.025 |
| Dabigatran | 472.00 | 288.50, 324.10 | 35.00 | 24.00 | 0.025 |
| [$^{13}C_6$]-Dabigatran | 478.00 | 295.00, 312.20 | 35.00 | 24.00 | 0.025 |

[a]Optimized $m/z$ may change based on tuning parameters

## 4 Notes

1. Mix samples vigorously and allow to sit at room temperature for at least 10 min to ensure adequate extraction.

2. Be sure that the supernatants are clear after centrifugation. If not, increase centrifugation time.

3. Be careful not to transfer any of the pellets from the bottom of the tubes following microcentrifugation.

### References

1. Gonsalves WI, Pruthi RK, Patnaik MM (2013) The new oral anticoagulants in clinical practice. Mayo Clin Proc 88:495–511

2. Stangier J (2008) Clinical pharmacokinetics and pharmacodynamics of the oral direct thrombin inhibitor dabigatran etexilate. Clin Pharmacokinet 47:285–295

3. Perzborn E, Roehrig S, Straub A et al (2010) Rivaroxaban: a new oral factor Xa inhibitor. Arterioscler Thromb Vasc Biol 30:376–381

4. Raghavan N, Frost CE, Yu Z et al (2009) Apixaban metabolism and pharmacokinetics after oral administration to humans. Drug Metab Dispos 37:74–81

# Chapter 4

# Simultaneous Quantitation of Lamotrigine, Levetiracetam, 10-Hydroxycarbazepine, Topiramate, and Zonisamide in Serum Using HPLC-MS/MS

**Dean C. Carlow, Heng Shi, and Ryan C. Schofield**

## Abstract

Antiepileptic drugs (AEDs) are a diverse group of pharmacological agents used in the treatment of epileptic seizures. Over the past several decades some new AEDs, including lamotrigine (LTG), levetiracetam (LVA), oxcarbazepine (OXC), topiramate (TOP), and zonisamide (ZNS), have become widely used. This chapter describes a very simple and rapid liquid chromatography-tandem mass spectrometry method for simultaneous quantitation of LVA, ZNS, LTG, TOP, and MHD in human serum. The method requires a very small amount of serum (50 μL) for multiple drug measurements and has a total analysis time of 4 min that makes it well suited for routine clinical analysis of several drugs simultaneously. The imprecision (CVs) measured at various concentrations across the analytical measurement range (AMR) are less than 7 % for all analytes. The AMR for each analyte is as follows: LVA (1–100 μg/mL), ZNS (0.8–80 μg/mL), TOP (0.5–50 μg/mL), and 0.6–60 μg/mL for LTG and MHD.

**Key words** Lamotrigine, Levetiracetam, 10-Hydroxycarbazepine, Topiramate, Zonisamide, Mass spectrometry, Therapeutic drug monitoring, Antiepileptic drugs

## 1 Introduction

Antiepileptic drugs (AEDs) are a diverse group of pharmacological agents used in the treatment of epileptic seizures. AEDs suppress the rapid and excessive firing of neurons during seizures and prevent the spread of the seizure within the brain. The AEDs are among the most common medications for which therapeutic drug monitoring (TDM) is performed. The reasons include the fact that the AEDs have a narrow therapeutic range and the clinical response often corresponds better to the drug concentration than the dose and the drugs display a pronounced intraindividual variation in their pharmacokinetics [1–3]. In addition, physiological markers for AED clinical efficacy or toxicity are not immediately apparent and seizures occur at irregular intervals; treatment is therefore prophylactic which makes determining the optimal dose difficult on clinical grounds alone [1–3].

Uttam Garg (ed.), *Clinical Applications of Mass Spectrometry in Drug Analysis: Methods and Protocols*, Methods in Molecular Biology, vol. 1383, DOI 10.1007/978-1-4939-3252-8_4, © Springer Science+Business Media New York 2016

Over the past several decades some new AEDs, including lamotrigine (LTG), levetiracetam (LVA), oxcarbazepine (OXC), topiramate (TOP), and zonisamide (ZNS), have been approved and become widely used. Compared with the older AEDs the newer drugs generally have a lower rate of side effects, once- or twice-daily dosing and for some, fewer drug-drug interactions [4, 5]. TDM is performed less frequently with newer AEDs because they generally display less pharmacokinetic variation than the older drugs and there is less documentation of a correlation between serum concentrations and clinical effects to support its use [4, 5]. However, TDM may be warranted for newer AEDs for a number of reasons: (a) to determine the optimal concentration for the individual patient; (b) to monitor compliance; (c) serum concentrations of the drugs that are eliminated renally (including topiramate and levetiracetam) can change drastically based on age, renal function, and pregnancy; (d) some of the newer AEDs are highly metabolized (including zonisamide, oxcarbazepine, and lamotrigine) and therefore will display variable pharmacokinetics; (e) one of the newer AEDs is a prodrug (oxcarbazepine) and the active metabolite 10-hydroxycarbazepine (MHD) is monitored clinically; (f) many patients on the newer AEDs receive multiple agents simultaneously which may result in drug-drug interactions [1–7].

This chapter describes a very simple and rapid liquid chromatography-tandem mass spectrometry method for simultaneous quantification of LVA, ZNS, LTG, TOP, and MHD in human serum. The method requires a very small amount of serum (50 µL) for multiple drug measurements and has a total analysis time of 4 min that makes it well suited for routine clinical analysis of several drugs simultaneously.

## 2    Materials

### 2.1    Samples

Serum or plasma (heparin) samples are acceptable for this procedure. Samples are stable 1 week when refrigerated or 4 weeks when frozen at –20 °C.

### 2.2    Reagents

1. Bovine serum albumin (Sigma-Aldrich).

2. Drug-free serum (UTAK Laboratories).

3. Mobile phase A (2.5 mM ammonium formate in water containing 1 % methanol): Add 980 mL of water to a 1 L graduated cylinder. To the cylinder add 10 mL methanol and 10 mL of 250 mM ammonium formate. Decant into a 1 L HPLC solvent bottle, cap, and invert ten times. Degas for 5 min by sonication. The mobile phase is stable at room temperature, 18–24 °C, for 1 month.

4. Mobile phase B (2.5 mM ammonium formate in methanol): Add 990 mL of methanol to a 1 L graduated cylinder. To the

cylinder add 10 mL of 250 mM ammonium formate. Decant into a 1 L HPLC solvent bottle, cap, and invert ten times. Degas for 5 min by sonication. The mobile phase is stable at room temperature, 18–24 °C, for 1 month.

5. Extraction Solution (0.1 M zinc sulfate solution containing 0.1 % formic acid): Using an analytical balance weigh 14.3 g zinc sulfate heptahydrate and add to a 500 mL solvent bottle. Add 500 mL water and 500 μL formic acid. Cap and mix well. The solution is stable refrigerated, 2–8 °C, for 1 month.

6. Internal standard solution (clonazepam-$D_4$ in methanol, 10 μg/mL): Add 9 mL of methanol to a 20 mL amber vial. Then add 1 mL of clonazepam-$D_4$ stock (100 μg/mL) and mix well. The solution is stable for 6 months when stored at –20 °C.

7. Injection solvent (mobile phase A/mobile phase B, 20:1 v/v): Using a 500 mL graduated cylinder add 475 mL mobile phase A and 25 mL mobile phase B. Decant into a 1 L HPLC solvent bottle, cap and invert ten times. Degas for 5 min by sonication. The solution is stable refrigerated, 2–8 °C, for 3 months.

8. Wash Solvent (methanol/2-propanol/water, 7:2:1 v/v): In a 500 mL graduated cylinder, add 350 mL of methanol, 100 mL of 2-propanol, and 50 mL of water and transfer into an HPLC wash bottle. Degas the solution for 5 min by sonication. The solution is stable at room temperature, 18–24 °C, up to 1 month.

*2.3  Internal Standards and Standards*

1. Primary standards: LVA (UCB Pharma), ZNS (Sigma-Aldrich), LTG (Sigma-Aldrich), TOP (Sigma-Aldrich), and MHD (Novartis).

2. Primary internal standard (I.S.): Clonazepam-$D_4$ 100 μg/mL (Sigma-Aldrich).

3. Standard stock solutions:

   (a) Levetiracetam (LVA, 800 μg/mL): Using an analytical balance weigh 20 mg of LVA and transfer into a 25 mL volumetric flask. Bring to volume with methanol and mix thoroughly to ensure homogeneity. Stock solution is stable for 1 year when stored at –80 °C.

   (b) Zonisamide (ZNS, 400 μg/mL): Using an analytical balance weigh 10 mg of ZNS and transfer into a 25 mL volumetric flask. Bring to volume with methanol and mix thoroughly to ensure homogeneity. Stock solution is stable for 1 year when stored at –80 °C.

   (c) Lamotrigine (LTG, 400 μg/mL): Using an analytical balance weigh 10 mg of LTG and transfer into a 25 mL volumetric flask. Bring to volume with methanol and mix thoroughly to ensure homogeneity. Stock solution is stable for 1 year when stored at –80 °C.

**Table 1**
**Calibrator concentrations (μg/mL)**

| Compound | LVA | ZNS | LTG | MHD | TOP |
|---|---|---|---|---|---|
| Calibrator 1 | 2.0 | 1.6 | 0.8 | 1.2 | 1.0 |
| Calibrator 2 | 10.0 | 8.0 | 4.0 | 6.0 | 5.0 |
| Calibrator 3 | 25.0 | 20.0 | 10.0 | 15.0 | 12.5 |
| Calibrator 4 | 50.0 | 40.0 | 20.0 | 30.0 | 25.0 |
| Calibrator 5 | 75.0 | 60.0 | 30.0 | 45.0 | 37.5 |
| Calibrator 6 | 100.0 | 80.0 | 40.0 | 60.0 | 50.0 |

(d) 10-Hydroxycarbazepine (MHD, 400 μg/mL): Using an analytical balance weigh 10 mg of MHD and transfer into a 25 mL volumetric flask. Bring to volume with methanol and mix thoroughly to ensure homogeneity. Stock solution is stable for 1 year when stored at –80 °C.

(e) Topiramate (TOP, 400 μg/mL): Using an analytical balance weigh 10 mg of TOP and transfer into a 25 mL volumetric flask. Bring to volume with methanol and mix thoroughly to ensure homogeneity. Stock solution is stable for 1 year when stored at –80 °C.

**2.4 Calibrators and Controls**

1. Calibrators: Six calibrators are used when generating a standard calibration curve with concentrations depicted in Table 1 (*see* **Notes 1** and **2**).

2. High calibrator preparation: Begin by making the high calibrator (calibrator 6) as follows: To three individual 13 × 100 mm glass tubes add LVA (250 μL), ZNS (400 μL), LTG (200 μL), MHD (300 μL), and TOP (250 μL) from their respective standard stock solutions. Evaporate the solvent of the high calibrator to dryness at 37 °C in a TurboVap evaporator using a stream of nitrogen. Reconstitute the compounds with 2 mL drug-free serum per tube and vortex mix thoroughly. Combine the three reconstituted serum tubes to generate 6 mL of calibrator 6. To produce calibrators 1–5 (*see* Table 2 and **Note 3**).

3. Check the new lot of standards by verifying five unknown patient samples with the current lot of calibrators. The agreement between the two calculated concentrations must be within 10 %.

4. Controls: AED II Serum Toxicology Controls (Bi-Level) were purchased from UTAK Laboratories. Reconstitute controls with 5 mL of water. The reconstituted control is stable for 25 days at 2–8 °C (*see* **Note 3**).

**Table 2**
**Calibrator preparation**

| Calibrator | Calibrator stock used | Volume added (µL) | Blank serum added (µL) |
|---|---|---|---|
| 1 | Calibrator 6 | 40 | 1960 |
| 2 | Calibrator 6 | 200 | 1800 |
| 3 | Calibrator 6 | 500 | 1500 |
| 4 | Calibrator 6 | 1000 | 1000 |
| 5 | Calibrator 6 | 1500 | 500 |
| Blank | 0 | 0 | 2000 |

5. Establish a range for the new lot of controls by collecting data points over 20 consecutive runs and establish the mean and standard deviation.

### 2.5 Analytical Equipment and Supplies

1. HPLC series 200 system (Perkin Elmer) coupled to an Applied Biosystems API 4000 triple quadrupole mass spectrometer (AB Sciex).
2. Guard Column: $C_{18}$, $2 \times 4$ mm (Phenomenex).
3. HPLC column: Gemini $C_{18}$, $50 \times 2.0$ mm i.d., 3 µm particle (Phenomenex).
4. Eppendorf 1.5 mL microcentrifuge tubes and National Scientific 2 mL amber glass vials with inserts and pre-slit caps or equivalent.
5. TurboVap LV evaporator (Biotage).

### 2.6 Instrument Operating Conditions

1. High-performance liquid chromatography (HPLC): A Perkin Elmer HPLC Series 200 system consisted of an autosampler, column oven, and two micro pumps. Chromatographic separations of the five AEDs and the internal standard were achieved using a $50 \times 2.0$ mm i.d., 3 µm particle size Gemini $C_{18}$ column, with a $C_{18}$ guard column ($2 \times 4$ mm) maintained at 50 °C. Mobile phase A consisted of 1 % methanol in water and 2.5 mM ammonium formate, and mobile phase B 2.5 mM ammonium formate in methanol. The HPLC method is described in Table 3. The injection volume is 5 µL, with a syringe wash volume of 250 µL using the syringe wash solvent. The autosampler performs one wash pre-injection and three washes post-injection.
2. Tandem mass spectrometry: Mass spectrometric detection was performed using an Applied Biosystems API 4000 triple quadropole mass spectrometer equipped with an electrospray

**Table 3**
**HPLC method**

| Step | Total time (min) | Flow rate (µL/min) | %A | %B |
|------|------------------|---------------------|-----|-----|
| 0 | 1.0 | 300 | 95 | 5 |
| 1 | 0.5 | 300 | 95 | 5 |
| 2 | 2.0 | 300 | 5 | 95 |
| 3 | 4.0 | 300 | 5 | 95 |

**Table 4**
**Analyte precursor and product ions (*m/z*)**

| Analyte | Precursor ion | 1° Product ion | 2° Product ion | CE (V) |
|---------|---------------|----------------|----------------|--------|
| LVA | 171 | 126 | 154 | 17 |
| ZNS | 213 | 132 | 77 | 22 |
| LTG | 256 | 211 | 145 | 38 |
| MHD | 255 | 194 | 237 | 25 |
| TOP | 340 | 264 | 282 | 25 |
| Clonazepam-$D_4$ | 320 | 274 | N/A | 35 |

Optimized *m/z* may change based on instrument and tuning parameters

ionization (ESI) source operating in a positive ion mode. Multiple reaction monitoring (MRM) was selected for detection of the five drugs and internal standard with a dwell time of 75 ms. As shown in Table 4, two mass transitions were monitored for each analyte. The tune parameters used for data acquisition were: source temperature of 300 °C; collision activation dissociation (CAD) gas value of 4; curtain gas of 30 psi; nebulizer gas of 35 psi; heating gas of 35 psi; and a spray voltage of 4500 V. Nitrogen (99.995 % purity) was used as the desolvation and collision gas. The MRM acquisition method was run in unit resolution (0.7 amu) in both Q1 and Q3.

## 3   Methods

### 3.1   Stepwise Procedure

1. Run a system suitability to confirm the system performance (*see* **Note 2**).

2. To 1.5 mL microcentrifuge tubes pipette 25 µL sample (calibrators, controls, or patient specimen) (*see* **Note 3**).

3. Add 50 µL of the extraction solution (0.1 M zinc sulfate solution with 0.1 % formic acid).

4. Add 50 µL of the internal standard solution (clonazepam-$D_4$ in methanol, 10 µg/mL).

5. Add 1 mL of methanol.

6. Cap and vortex each sample at maximum speed for 5 s.

7. Centrifuge for 5 min at $13,800 \times g$.

8. Transfer 1 mL of the injection solvent to an appropriately labeled glass autosampler vial.

9. Transfer 50 µL of the supernatant to 1 mL of the injection solvent.

10. Place all samples in the autosampler and inject 5 µL of the sample into the LC-ESI-MS/MS (*see* **Note 4**).

*3.2 Analysis*

1. The data are analyzed using Analyst 1.4.1 software (AB Sciex).

2. Standard curves are based on a linear regression for all analytes. Weighted linear regression models with weights inversely proportional to the *X* values were used. The analysis compared I.S. peak area to sample peak area (*y*-axis) versus analyte concentration (*x*-axis) using the quantifying ions indicated in Table 4.

3. Acceptability of each run is confirmed if the calculated control concentrations fall within 2 standard deviations of the target mean values. Target values are established as the mean of 20 runs. If any of the controls are greater than 3 standard deviations the run cannot proceed and troubleshooting procedure must commence.

4. Typical coefficients of correlation of the standard curve are >0.99 (*see* **Note 5**).

5. Typical chromatograms for a calibrator and a patient sample are shown in Figs. 1 and 2.

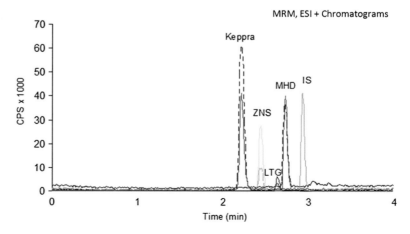

**Fig. 1** LC-ESI-MS/MS ion chromatograms of LVA, ZNS, LTG, MHD, and clonazepam-D4 (I.S.) primary product ions from a calibrator sample

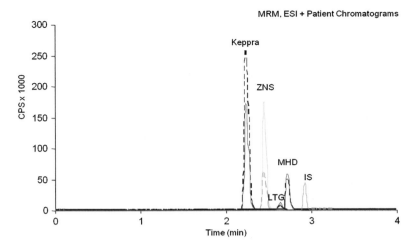

**Fig. 2** LC-ESI-MS/MS ion chromatograms of LVA, ZNS, LTG, MHD, and clonaze-pam-D4 (I.S.) primary product ions from a patient sample

# 4  Notes

1. A new standard curve should be generated with each analytical run to ensure method performance.

2. A system suitability should be performed each day the method is run. The suitability includes running a test mix with all analytes to ensure proper retention time, integration, and sensitivity.

3. Individual sets of calibrators and controls can be aliquoted and stored at −80 °C for 1 year. For each calibrator or control, aliquot 75 μL into a glass inset and place in a 2 mL amber glass vial and cap. Thaw completely before use.

4. Retention times are instrument specific and can vary due to column use and PEEK tubing length.

5. The imprecision (CVs) measured at various concentrations across the analytical measurement range (AMR) are less than 7 % for all analytes. The calibration curves are linear over the AMR with correlation coefficients $r \geq 0.99$. The AMR for each analyte is as follows: LVA (1–100 μg/mL), ZNS (0.8–80 μg/mL), TOP (0.5–50 μg/mL), and 0.6–60 μg/mL for LTG and MHD.

## References

1. Schachter SC (1999) Antiepileptic drug therapy: general treatment principles and application for special patient populations. Epilepsia 40(s9):S20–S25

2. Neels HM et al (2004) Therapeutic drug monitoring of old and newer anti-epileptic drugs. Clin Chem Lab Med 42(11):1228–1255

3. Patsalos PN et al (2008) Antiepileptic drugs—best practice guidelines for therapeutic drug monitoring: a position paper by the subcommission on therapeutic drug monitoring, ILAE Commission on Therapeutic Strategies. Epilepsia 49(7):1239–1276

4. Johannessen SI et al (2003) Therapeutic drug monitoring of the newer antiepileptic drugs. Ther Drug Monit 25(3):347–363

5. Johannessen SI, Tomson T (2006) Pharmacokinetic variability of newer antiepileptic drugs. Clin Pharmacokinet 45(11):1061–1075

6. Perucca E (2000) Is there a role for therapeutic drug monitoring of new anticonvulsants. Clin Pharmacokinet 38(3):191–204

7. Krasowski MD (2010) Therapeutic drug monitoring of the newer anti-epilepsy medications. Pharmaceuticals 3(6):1909–1935

# Chapter 5

# Quantification of the Triazole Antifungal Compounds Voriconazole and Posaconazole in Human Serum or Plasma Using Liquid Chromatography Electrospray Tandem Mass Spectrometry (HPLC-ESI-MS/MS)

## Alejandro R. Molinelli and Charles H. Rose IV

### Abstract

Voriconazole and posaconazole are triazole antifungal compounds used in the treatment of fungal infections. Therapeutic drug monitoring of both compounds is recommended in order to guide drug dosing to achieve optimal blood concentrations. In this chapter we describe an HPLC-ESI-MS/MS method for the quantification of both compounds in human plasma or serum following a simple specimen preparation procedure. Specimen preparation consists of protein precipitation using methanol and acetonitrile followed by a cleanup step that involves filtration through a cellulose acetate membrane. The specimen is then injected into an HPLC-ESI-MS/MS equipped with a C18 column and separated over an acetonitrile gradient. Quantification of the drugs in the specimen is achieved by comparing the response of the unknown specimen to that of the calibrators in the standard curve using multiple reaction monitoring.

**Key words** Antifungal, Triazole, Voriconazole, Posaconazole, Mass spectrometry, Fungal infection, Therapeutic drug monitoring

## 1 Introduction

Voriconazole and posaconazole are second-generation triazole antifungal compounds used in the management of fungal infections. The mechanism of action for both drugs involves blocking the conversion of lanosterol to ergosterol in the fungal cell membrane by inhibition of the cytochrome P450 enzyme 14α-lanosterol demethylase [1, 2]. Voriconazole is a broad-spectrum antifungal agent that exhibits activity against molds, yeasts, and endemic mycoses; it is available in oral and parenteral formulations. Voriconazole is metabolized in the liver by CYP2C19, CYP3A4, and CYP2C9 [3]. The presence of CYP2C19 gain-of-function alleles in particular is associated with voriconazole concentrations that are below the therapeutic range in children [4]. A prospective,

Uttam Garg (ed.), *Clinical Applications of Mass Spectrometry in Drug Analysis: Methods and Protocols*, Methods in Molecular Biology, vol. 1383, DOI 10.1007/978-1-4939-3252-8_5, © Springer Science+Business Media New York 2016

randomized controlled trial in adults found that outcomes in patients undergoing therapeutic drug monitoring of voriconazole were significantly better than those in patients who received a fixed regimen of the drug [5].

Posaconazole has the widest spectrum of antifungal activity of any triazole and is the first to demonstrate activity against zygomycetes. It is indicated for the prophylaxis of invasive *Aspergillus* spp. and *Candida* spp. infections in severely immunocompromised patients [6]. Posaconazole absorption is affected by fatty foods, gastric pH, and mucosal health. Posaconazole bioavailability is also affected by its saturable absorption [3]. Therapeutic drug monitoring of posaconazole is recommended given its pharmacokinetic variability, the relationships between concentration and effect observed in experimental models of invasive fungal infections, and findings of sub-therapeutic concentrations in patients receiving fixed regimens of the drug [6, 7].

Triazole compounds can be monitored using various analytical techniques including HPLC-UV and LC-MS/MS [8–12]. The method described here involves the quantification of voriconazole and posaconazole following a simple sample preparation procedure. Sample preparation consists of protein precipitation using methanol and acetonitrile followed by a cleanup step using cellulose acetate filters. The specimen is then injected into an HPLC-ESI-MS/MS equipped with a C18 column and separated over an acetonitrile gradient. The detector consists of a tandem mass spectrometer operated in the multiple reaction monitoring mode. The precursor and two product ions are monitored per analyte.

## 2  Materials

### 2.1  Samples

Human plasma (EDTA) or serum is acceptable for this procedure. Samples are stable for 1 month when refrigerated at 4 °C or 1 year when frozen at –20 °C.

### 2.2  Reagents

1. Ammonium formate buffer (20 mM, pH 3.8): Add 1.26 g of ammonium formate to 995 mL of deionized water contained in a 1 L glass beaker. Mix using a magnetic stirrer. Adjust the pH to 3.8 with formic acid or sodium hydroxide and adjust the volume to 1 L after transferring to a volumetric flask. Recheck the pH and adjust if necessary. Store the buffer in a capped bottle at 4 °C. Use within 1 month.

2. Protein precipitation solution (methanol/acetonitrile (1:1) containing 0.1 % formic acid): Combine 50 mL of LC/MS-grade methanol and 50 mL of LC/MS-grade acetonitrile in a 100 mL glass bottle. Add 100 μL of formic acid to 100 mL of methanol/acetonitrile 1:1 (v/v). Store at room temperature

for up to 6 months. Keep lid tightly closed to prevent evaporation.

3. Methanol 75 %: Add 12.5 mL of deionized water to 37.5 mL of LC/MS-grade methanol and mix thoroughly. Store at room temperature. Use within 6 months.

4. Mobile phase A (40 % acetonitrile/2 mM ammonium formate buffer): Add 200 mL of LC/MS grade acetonitrile, and 50 mL of 20 mM ammonium formate to 250 mL deionized water. Mix the solution well and filter under vacuum through a 0.2 μm nylon membrane filter. Store in a screw-capped bottle. Stable for 1 week at room temperature.

5. Mobile phase B (70 % acetonitrile/2 mM ammonium formate buffer): Add 350 mL of LC/MS grade acetonitrile, and 50 mL of 20 mM ammonium formate to 100 mL of deionized water. Mix the solution well and filter under vacuum through a 0.2 μm nylon membrane filter. Store in a screw-capped bottle. Stable for 1 week at room temperature.

6. Strong wash solution (90 % acetonitrile/methanol 1:1): Combine 450 mL of LC/MS-grade acetonitrile, 450 mL of LC/MS-grade methanol, and 100 mL of deionized water. Mix the solution well and filter under vacuum through a 0.2 μm nylon membrane filter. Stable for 6 months at room temperature.

7. Weak wash solution (10 % acetonitrile/methanol 1:1): Combine 50 mL of LC/MS-grade acetonitrile, 50 mL LC/MS-grade methanol, and 900 mL of deionized water. Mix the solution well and filter under vacuum through a 0.2 μm nylon membrane filter. Stable for 6 months at room temperature.

8. Column rinse solution (80 % acetonitrile): Combine 800 mL of LC/MS-grade acetonitrile with 200 mL deionized water. Mix the solution well and filter under vacuum through a 0.2 μm nylon membrane filter. Stable for 6 months at room temperature.

### 2.3 Standards and Calibrators

1. Working calibrators: Lyophilized human serum calibrators containing voriconazole and posaconazole are obtained from a commercial source (UTAK Laboratories, Inc., Valencia, CA) at the following concentrations: 0.025, 0.1, 1.0, and 20.0 μg/mL.

2. The working calibrators should be reconstituted according to the manufacturer's instructions prior to use.

3. Lyophilized product is stable for 2 years at 2–8 °C. Reconstituted material is stable for 25 days at 2–8 °C.

### 2.4 Quality Controls and Internal Standard

1. Quality controls: Lyophilized human serum quality control samples containing voriconazole and posaconazole are obtained from a commercial source (UTAK Laboratories, Inc.,

Valencia, CA) at the following nominal concentrations: level 1 at 0.5 μg/mL; level 2 at 5.0 μg/mL.

2. The quality control samples should be reconstituted according to the manufacturer's instructions prior to use.

3. Lyophilized product is stable for 2 years at 2–8 °C. Reconstituted material is stable for 25 days at 2–8 °C.

4. Stock internal standard: 1 mg/mL Voriconazole-d3 (Toronto Research Chemicals, Toronto, Canada). Prepare by adding 1 mL of 75 % methanol to 1 mg of solid voriconazole-d3.

5. Internal standard primary solutions: Add 50 μL of stock internal standard to a 50 mL volumetric flask and bring the volume to the mark with the 75 % methanol solution. This will yield a 1 μg/mL solution. Label this solution "working solution A." Add 0.5 mL of "working solution A" to a 5 mL volumetric flask and bring the volume to the mark with the 75 % methanol solution. This will yield a 100 ng/mL solution. Label this solution "working solution B."

6. Internal standard working solution: Add 1.5 mL of "working solution B" to a 5 mL volumetric flask and bring the volume to the mark with the 75 % methanol solution. This will yield a 30 ng/mL working internal standard. Stable for 1 year when stored at 2–8 °C.

**2.5 Supplies and Equipment**

1. Glass bottles with screw caps (1 L, 500 mL, and 250 mL capacities) (*see* **Note 1**).

2. Vacuum filtering apparatus fitted with 0.2 μm nylon membrane.

3. A liquid chromatograph-electrospray ionization tandem-mass spectrometer with autosampler operated in the multiple reaction monitoring mode (Waters Acquity UPLC with a Xevo TQ-MS mass spectrometer, using MassLynx software, Waters Corporation, Milford, MA).

4. Analytical column: Waters Cortecs C18 RP UPLC column (2.1 mm I.D. × 50 mm, 1.6 μm).

5. Waters Guard Column (2.1 mm × 0.2 μm) and frits.

6. Waters Certified UPLC autosampler vials (12 mm × 32 mm) (high recovery).

# 3 Methods

**3.1 Stepwise Procedure**

1. Add 100 μL of specimen (calibrator, quality control, unknown (plasma or serum), or blank matrix) to a labeled 1.5 mL microcentrifuge tube.

2. Add 20 μL internal standard working solution (voriconazole-d3, 30 ng/mL) to each tube.

3. Cap tubes and vortex mix at maximum speed for 5 s.

4. Add 300 µL of protein precipitation solution to each tube.

5. Cap tubes and vortex mix at maximum speed for 15 s.

6. Centrifuge tubes for 5 min at $16,000 \times g$.

7. Transfer 100 µL of the supernatant to labeled 1.5 mL micro-centrifuge tubes fitted with the 0.2 µm cellulose acetate filters.

8. Centrifuge tubes for 3 min at $16,000 \times g$.

9. Transfer filtrate to labeled autosampler vials and place the vials in the autosampler.

10. Inject 5 µL of sample onto HPLC-ESI-MS/MS. Sample ion chromatograms are shown in Fig. 1.

### 3.2 Data Analysis

1. Instrument operating parameters are given in Table 1a, b.

2. Data are analyzed using TargetLynx application of the MassLynx software (Waters Inc.).

3. The quantifying ions (Table 2) are used to construct standard curves of the peak area ratios (calibrator/internal standard pair) ($y$) versus analyte concentration ($x$). These curves are then used to determine the concentrations of the quality control and unknown (patient) specimens.

4. Representative LC-ESI-MS/MS chromatograms of voriconazole, posaconazole, and voriconazole-d3 are shown in Fig. 1a, b.

5. Acceptability of each run is dependent on proper system suitability parameters (*see* **Notes 2** and **3**), calibration curve, and quality control specimen performance. A typical calibration curve has a coefficient of determination ($R^2$) of >0.99.

6. Quality control specimens should fall within the acceptable parameters established by each lab (e.g., using Westgard Rules). Target values for the commercial controls described in Subheading 2.4 are established in-house after a minimum of 20 runs.

7. The quantifying ion in the sample is considered acceptable if the ratios of qualifier ions to quantifying ion are within 20 % of the ion ratios for the calibrators.

8. Liquid chromatography retention time window limits are set at ±0.25 min. Typical retention times are 0.55, 0.59, and 0.98 min for voriconazole, voriconazole d-3, and posaconazole, respectively (*see* **Note 4**).

9. The performance characteristics of the assay are presented in Table 3.

10. Ion suppression effects were evaluated by the sample infusion method and found to be non-significant. However, each lab should evaluate ion suppression independently.

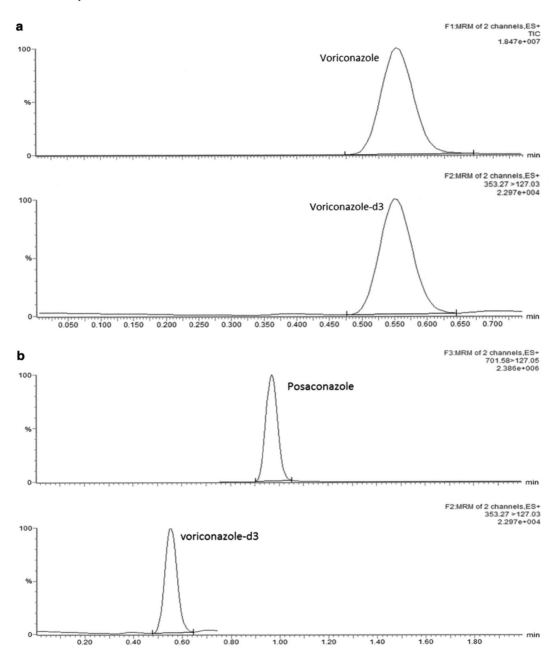

**Fig. 1** (**a**) HPLC-ESI-MS/MS ion chromatograms of voriconazole and voriconazole-d3. Only the primary ions are shown. (**b**) HPLC-ESI-MS/MS ion chromatograms of posaconazole and voriconazole-d3. Only the primary ions are shown

**Table 1**
**HPLC-ESI-MS/MS operating conditions[a]**

| A. Liquid chromatography gradient conditions | | | |
|---|---|---|---|
| Time (min) | Flow rate (mL/min) | Mobile phase A (%) | Mobile phase B (%) |
| 0 | 0.35 | 60 | 40 |
| 1.5 | 0.35 | 0 | 100 |
| 1.75 | 0.35 | 0 | 100 |
| 2.05 | 0.35 | 60 | 40 |
| 2.5 | 0.35 | 60 | 40 |
| B. MS/MS tune settings | | | |
| Capillary (kV) | 2.9 | | |
| Cone (V) | 29 | | |
| Source temperature (°C) | 150 | | |
| Desolvation temperature (°C) | 350 | | |
| Cone gas flow (L/h) | 1.0 | | |
| Desolvation gas flow (L/h) | 650 | | |
| Resolution MS1 and MS2 | MS1 = 1.0 MS2 = 0.75 | | |

[a]Conditions optimized for instrumentation described in Subheading 2.5. MS/MS tune settings may vary between instruments

**Table 2**
**Precursor and primary and secondary product ions for voriconazole, posaconazole, and voriconazole-d3**

| | Precursor ion $(M+H)^{+1}$ | Product ions (primary, secondary) | Cone (V) | Collision (eV) |
|---|---|---|---|---|
| Voriconazole | 350.08 | 126.97, 281.10 | 24 | 32, 16 |
| Voriconazole-d3 | 353.08 | 127.03, 284.10 | 24 | 40, 16 |
| Posaconazole | 701.39 | 127.05, 614.34 | 58 | 76, 36 |

**Table 3**
**Method performance specifications**

|                                  | Voriconazole                                      | Posaconazole                                       |
| -------------------------------- | ------------------------------------------------- | -------------------------------------------------- |
| Analytical measurement range     | 0.025–20.0 µg/mL                                  | 0.025–20.0 µg/mL                                   |
| Linearity                        | $m=0.988$, $b=0.0674$, $r^2=0.9997$               | $m=1.01$, $b=-0.0054$, $r^2=0.9996$                |
| [a]Intra-assay precision (CV%)   | L1 = 8.7 %, L2 = 3.4 %, L3 = 4.9 %                | L1 = 9.2 %, L2 = 3.2 %, L3 = 5.3 %                 |
| [a]Inter-assay precision (CV%)   | L1 = 9.8 %, L2 = 5.7 %, L3 = 7.1 %                | L1 = 13.0 %, L2 = 5.2 %, L3 = 6.7 %                |
| [b]LoD, LoQ (µg/mL)              | LoD = 0.005, LoQ = 0.025                           | LoD = 0.005, LoQ = 0.025                           |
| Recovery                         | 97.60 %                                           | 95.10 %                                            |

[a]L1 = level 1 (0.075 µg/mL); L2 = level 2 (5.00 µg/mL); L3 = level 3 (10.00 µg/mL)
[b]LoD = limit of detection (CLSI EP17-A2); LoQ = limit of quantitation chosen as lowest calibrator value (CV < 20 %)

## 4   Notes

1. Stock reagent solutions: All reagent glassware (bottles, flasks, cylinders, or containers) should be properly labeled with the name of reagent, the date prepared, the date of expiration, and the initials of the analyst who prepared the reagent. Do not use glassware that has been exposed to detergents for this assay. Detergents can contaminate the mass spectrometer. Be careful to only use glassware that has been properly rinsed with methanol or acetonitrile (LC/MS grade) and allowed to dry before usage.

2. A tuning solution (used to tune the instrument prior to an analytical run) consists of a mixture of voriconazole, voriconazole d-3, and posaconazole, each diluted to 1 µg/mL in 75 % methanol (25 mL final volume).

3. A system suitability specimen (check standard) is tested with each run to verify instrument function (e.g., peak shape, retention time).

4. Retention times are instrumentation specific and can vary from what is described here.

## References

1. Product information (2010) VFEND® (voriconazole). Available via Drugs@FDA database. http://www.accessdata.fda.gov/drugsatfda_docs/label/2010/021266s032lbl.pdf. Accessed 25 Feb 2015

2. Product information (2013) Noxafil® (posaconazole). Available via Drugs@FDA database. http://www.accessdata.fda.gov/drugsatfda_docs/label/2013/205053s000lbl.pdf. Accessed 25 Feb 2015

3. Andes D, Pascual A, Marchetti O (2009) Antifungal therapeutic drug monitoring: established and emerging indications. Antimicrob Agents Chemother 53(1):24–34

4. Hicks JK, Crews KR, Flynn P, Haidar CE, Daniels CC, Yang W, Panetta JC, Pei D, Scott

JR, Molinelli AR, Broeckel U, Bhojwani D, Evans WE, Relling MV (2014) Voriconazole plasma concentrations in immunocompromised pediatric patients vary by CYP2C19 diplotypes. Pharmacogenomics 15(8):1065–1078

5. Park WB, Kim NH, Kim KH, Lee SH, Nam WS, Yoon SH, Song KH, Choe PG, Kim NJ, Jang IJ, Oh MD, Yu KS (2012) The effect of therapeutic drug monitoring on safety and efficacy of voriconazole in invasive fungal infections: a randomized controlled trial. Clin Infect Dis 55(8):1080–1087

6. Ashbee HR, Barnes RA, Johnson EM, Richardson MD, Gorton R, Hope WW (2014) Therapeutic drug monitoring (TDM) of antifungal agents: guidelines from the British Society for Medical Mycology. J Antimicrob Chemother 69(5):1162–1176

7. Bernardo VA, Cross SJ, Crews KR, Flynn PM, Hoffman JM, Knapp KM, Pauley JL, Molinelli AR, Greene WL (2013) Posaconazole therapeutic drug monitoring in pediatric patients and young adults with cancer. Ann Pharmacother 47(7–8):976–983

8. Pennick GJ, Clark M, Sutton DA, Rinaldi MG (2003) Development and validation of a high-performance liquid chromatography assay for voriconazole. Antimicrob Agents Chemother 47(7):2348–2350

9. Langman LJ, Boakye-Agyeman F (2007) Measurement of voriconazole in serum and plasma. Clin Biochem 40(18):1378–1385

10. Gordien JB, Pigneux A, Vigouroux S, Tabrizi R, Accoceberry I, Bernadou JM, Rouault A, Saux MC, Breilh D (2009) Simultaneous determination of five systemic azoles in plasma by high-performance liquid chromatography with ultraviolet detection. J Pharm Biomed Anal 50(5):932–938

11. Pauwels S, Vermeersch P, Van Eldere J, Desmet K (2012) Fast and simple LC-MS/MS method for quantifying plasma voriconazole. Clin Chim Acta 413(7–8):740–743

12. Beste KY, Burkhardt O, Kaever V (2012) Rapid HPLC-MS/MS method for simultaneous quantitation of four routinely administered triazole antifungals in human plasma. Clin Chim Acta 413(1–2):240–245

# Chapter 6

# Quantitation of Haloperidol, Fluphenazine, Perphenazine, and Thiothixene in Serum or Plasma Using Liquid Chromatography-Tandem Mass Spectrometry (LC-MS/MS)

## Matthew H. Slawson and Kamisha L. Johnson-Davis

### Abstract

Haloperidol, fluphenazine, perphenazine, and thiothixene are "typical" antipsychotic drugs that are used in the treatment of schizophrenia and other psychiatric disorders. The monitoring of the use of these drugs has applications in therapeutic drug monitoring and overdose situations. LC-MS/MS is used to analyze plasma/serum extracts with deuterated analog of imipramine as the internal standard to ensure accurate quantitation and control for any potential matrix effects. Positive ion electrospray is used to introduce the analytes into the mass spectrometer. Selected reaction monitoring of two product ions for each analyte allows for the calculation of ion ratios which ensures correct identification of each analyte, while a matrix-matched calibration curve is used for quantitation.

**Key words** Haloperidol, Fluphenazine, Perphenazine, Thiothixene, Plasma, Serum, UPLC, Mass spectrometry

## 1 Introduction

Fluphenazine (e.g. Prolixin) and perphenazine (e.g. Etrafon) are phenothiazine neuroleptics used in the management of psychotic disorders, such as schizophrenia, mania, anxiety/agitation, and depression. Haloperidol (e.g. Haldol) is a butyrophenone typical antipsychotic drug indicated for use in the treatment of schizophrenia and the control of tics and vocal utterances of Tourette's disorder in children and adults. Thiothixene (e.g. Navane) is a thioxanthene neuroleptic with general properties similar to those of the phenothiazines. Therapeutic monitoring of concentrations of these drugs is useful in optimizing therapy, evaluate compliance, and to monitor for adverse drug reaction [1, 2].

This chapter describes an analytical method to measure the above-mentioned four typical antipsychotic drugs in human serum/plasma by precipitating serum/plasma proteins and

Uttam Garg (ed.), *Clinical Applications of Mass Spectrometry in Drug Analysis: Methods and Protocols*, Methods in Molecular Biology, vol. 1383, DOI 10.1007/978-1-4939-3252-8_6, © Springer Science+Business Media New York 2016

collecting the supernatant for analysis. The supernatant is injected onto the LC-MS/MS. Qualitative identification is made using unique MS/MS transitions, ion ratios of those transitions, and chromatographic retention time. Quantitation is performed using a daily calibration curve of prepared calibration samples and using peak area ratios of analyte to internal standard to establish the calibration model. Patient sample concentrations are calculated based on the calibration model's mathematical equation. Quantitative accuracy is monitored with QC samples independently prepared with known concentrations of analyte and comparing the calculated concentration with the expected concentration [3, 4].

## 2    Materials

### 2.1    Samples

1. Pre-dose (trough) draw—at steady-state concentration for serum/plasma. Separate serum or plasma from cells within 2 h of collection.

2. Collect in plain red tube. Avoid gel or other separator tubes.

3. Specimens can be stored for at least 24 h ambient, 5 days refrigerated, 30 days frozen prior to analysis.

### 2.2    Reagents

1. Clinical Laboratory Reagent Water (CLRW).

2. Verified negative serum/plasma pool.

3. Mobile Phase A (CLRW with 0.1 % Formic Acid): 1.0 mL of concentrated Formic Acid in CLRW q.s. to 1.0 L in volumetric flask.

4. Mobile Phase B (Acetonitrile with 0.1 % Formic Acid): 1.0 mL of concentrated Formic Acid in LC-MS grade Acetonitrile q.s. to 1.0 L in volumetric flask.

### 2.3    Standards and Calibrators

1. Haloperidol, 1.0 mg/mL stock standard prepared in Methanol (Cerilliant, Round Rock, TX).

2. Perphenazine 1.0 mg/mL stock standard prepared in Methanol (Cerilliant, Round Rock, TX).

3. Fluphenazine 1.0 mg/mL stock standard prepared in Methanol (Cerilliant, Round Rock, TX).

4. Thiothixene 1.0 mg/mL stock standard prepared in Methanol (Cerilliant, Round Rock, TX).

5. Prepare an intermediate solution containing fluphenazine and perphenazine at 1000 ng/mL and haloperidol and thiothixene at 5000 ng/mL in methanol. Add ~3 mL of methanol to a 10 mL volumetric flask. Add 10 μL each of fluphenazine and perphenazine reference materials and 50 μL each of haloperidol and thiothixene reference material to the flask, q.s. to

**Table 1**
**Preparation of calibrators. The total volume is made to 25 mL with drug-free human serum/plasma**

| Calibrator | Volume of intermediate solution (µL) | Final [], ng/mL |
|---|---|---|
| 1 | 5 | 0.2/1[a] |
| 2 | 50 | 2/10[a] |
| 3 | 100 | 4/20[a] |
| 4 | 300 | 12/60[a] |

[a]Fluphenazine and perphenazine/haloperidol and thiothixene

10 mL with methanol, add a stir bar and stopper and mix for 30 min at room temperature. Aliquot as appropriate for subsequent use. Store frozen, stable for 1 year. This volume can be scaled up or down as appropriate.

6. Prepare working calibrators to prepare 25 mL of each using volumetric glassware. Add approximately 10 mL certified negative plasma/serum to a labeled volumetric flask. Add the appropriate volume as shown in Table 1 of intermediate solution described in **item 5** above to the flask; q.s. to 25 mL using certified negative serum/plasma. Add a stir bar and stopper and mix for at least 30 min at room temperature. Aliquot as appropriate for future use. Store aliquots frozen, stable for 1 year. This volume can be scaled up or down as appropriate (*see* **Note 1**).

**2.4 Controls and Internal Standard**

1. Controls: May be purchased from a third party and prepared according to the manufacturer. They can also prepared in-house independently from calibrators' source material using Table 1 as a guideline (*see* **Note 1**).

2. Internal Standard (protein precipitation solution): Imipramine-D$_3$ 100 mcg/mL in methanol (Cerilliant, Round Rock, TX). Add 250 mL of methanol to a 500 mL volumetric flask. Add 60 µL of reference material to the flask, QS to 500 mL with acetonitrile. Add a stir bar and a stopper. Mix for at least 30 min at room temperature. Aliquot as needed for use in this assay (volumes can be scaled up or down as appropriate). Store frozen, stable for 1 year (*see* **Notes 1** and **2**).

**2.5 Supplies and Equipment**

1. Instrument-compatible autosampler vials with injector appropriate caps.

2. Acquity HSS T3 1.8 µm, 2.1 × 50 mm UPLC column (Waters, Milford, MA).

3. Multi-tube Vortex mixer (e.g., VWR VX-2500).

4. Foam rack(s) compatible with both microcentrifuge tubes and multi-tube vortex mixer.

5. Centrifuge capable of $18,000 \times g$ that will accommodate microcentrifuge tubes.

6. Waters Acquity TQD UPLC-MS/MS system (Milford, MA).

# 3   Methods

## 3.1   Stepwise Procedure

1. Briefly vortex or invert each sample to mix.

2. Aliquot 100 μL of each patient sample, calibrator and QC into appropriately labeled microcentrifuge tubes.

3. Add 300 μL of Internal Standard/precipitation solution to each vial.

4. Cap each tube and vortex vigorously for 30 s.

5. Centrifuge for ~10 min at ~$18,000 \times g$ (*see* **Note 3**).

6. Transfer the contents of each tube (from **steps 2** to **5**) to an autosampler vial and cap.

7. Analyze on LC-MS/MS.

## 3.2   Instrument Operating Conditions

1. Table 2 summarizes typical LC conditions.

2. Table 3 summarizes typical MS conditions.

3. Table 4 summarizes typical MRM conditions.

Each instrument should be individually optimized for best method performance.

## 3.3   Data Analysis

1. Representative MRM chromatograms of each antipsychotic and internal standard in plasma are shown in Fig. 1a–e.

2. The dynamic range for this assay is 0.2–12 ng/mL for fluphenazine and perphenazine and 1–60 ng/mL for haloperidol and thiothixene. Samples exceeding this range can be diluted 5× or 10× as needed to achieve an accurate calculated concentration, if needed.

3. Data analysis is performed using the QuanLynx or TargetLynx software to integrate peaks, calculate peak area ratios, and construct calibration curves using a linear $1/x$ weighted fit ignoring the origin as a data point. Sample concentrations are then calculated using the derived calibration curves (*see* **Note 1**).

**Table 2**
**Typical HPLC conditions**

| Weak wash | Mobile Phase A | | | | |
|---|---|---|---|---|---|
| Strong wash | Mobile Phase B | | | | |
| Seal wash | Mobile Phase A | | | | |
| Injection volume | 8 µL | | | | |
| Vacuum degassing | On | | | | |
| Temperature | 30 °C | | | | |
| A Reservoir | 0.1 % HCOOH in CLRW | | | | |
| B Reservoir | 0.1 % HCOOH in Acetonitrile | | | | |
| Gradient table | | | | | |
| Step | Time (min) | Flow (µL/min) | A (%) | B (%) | Curve[a] |
| 0 | 0 | 650 | 70 | 30 | 1 |
| 1 | 1 | 650 | 55 | 45 | 6 |
| 2 | 1.33 | 650 | 10 | 90 | 6 |
| 3 | 1.55 | 650 | 70 | 30 | 11 |

[a]Nonlinear gradient curves common to Waters systems

**Table 3**
**Typical mass spectrometer conditions**

| Parameter | Value |
|---|---|
| Capillary (kV) | 0.6 |
| Cone (V) | 42 |
| Extractor (V) | 3 |
| RF (V) | 0.3 |
| Desolvation temp | 450 |
| Desolvation gas | 900 |
| Cone gas | 30 |
| Collision gas | 0.25 |
| Scan mode | MSMS |
| Polarity | Positive |
| Ion source | ESI |
| Resolution Q1 | Unit |
| Resolution Q3 | Unit |
| Dwell (s) | 0.045 |

**Table 4**
**Typical MRM conditions**

| Analyte | Precursor | Product (quant.) | Product (qual.) |
|---|---|---|---|
| Haloperidol | 376.2 | 165.1 | 122.9 |
|  | Collision energy | 22 | 42 |
| Fluphenazine | 438.3 | 171.1 | 143.1 |
|  | Collision energy | 26 | 32 |
| Perphenazine | 404.2 | 143.1 | 171.11 |
|  | Collision energy | 28 | 24 |
| Thiothixene | 444.3 | 139.2 | 97.9 |
|  | Collision energy | 34 | 34 |
| Imipramine-d$_3$ (internal standard) | 284.2 | 89.1 | 193.1 |
|  | Collision energy | 16 | 42 |

4. Calibration curves should have an $r^2$ value $\geq 0.99$.

5. Typical imprecision is <15 % both inter- and intra-assay.

6. An analytical batch is considered acceptable if chromatography is acceptable and QC samples calculate to within 20 % if their target values and ion ratios are within 20 % of the calibration curve ion ratios.

# 4    Notes

1. Validate/verify all calibrators, QCs, internal standard, and negative matrix pools before placing into use.

2. Imipramine-d$_3$ shows good recovery and a retention time intermediate to the other analytes making it a good compromise internal standard for all four antipsychotics. Ion suppression studies (data not shown) indicate that this I.S. offers good control of matrix effects under the conditions described. Deuterated analogs of each drug may be utilized if desired.

3. Time and speed of centrifugation step can be optimized to ensure a firm pellet is formed so as not to transfer any precipitate to autosampler vial.

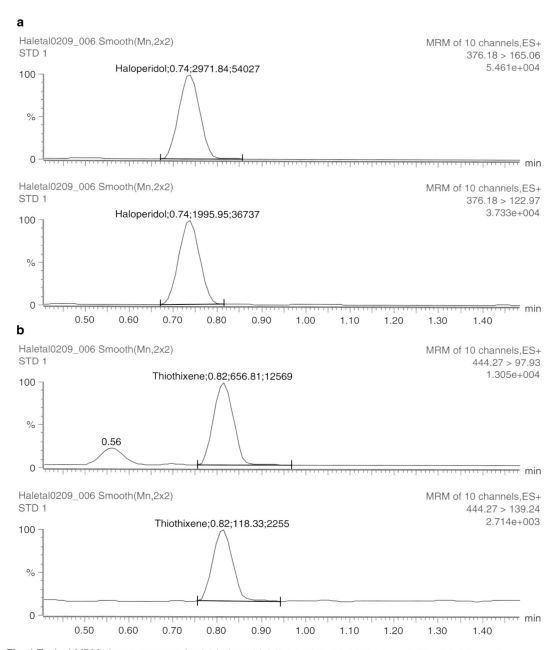

**Fig. 1** Typical MRM chromatograms for (**a**) haloperidol (0.74 min), (**b**) thiothixene (0.82 min), (**c**) perphenazine (0.84 min), (**d**) imipramine-d3 (0.88 min), (**e**) fluphenazine (1.01 min) in plasma. 0.2/1 ng/mL (see text) extracted from fortified human plasma and analyzed according to the described method

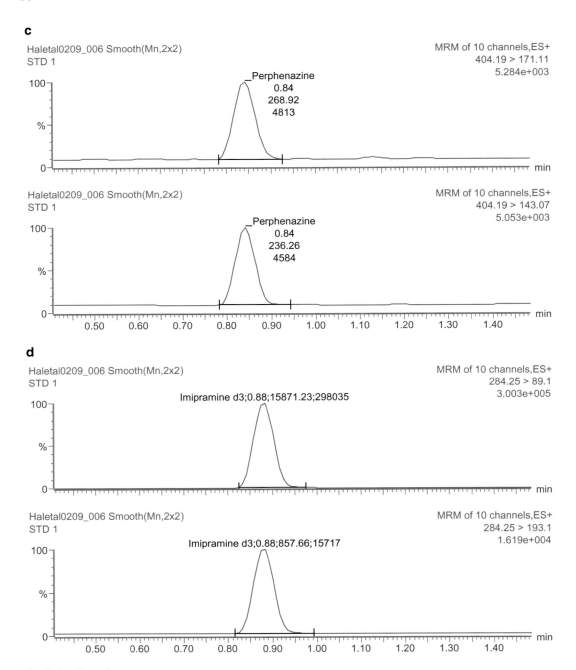

**Fig. 1** (continued)

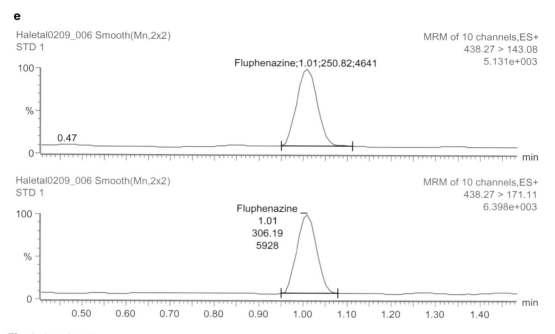

**Fig. 1** (continued)

## References

1. Baselt RC (2011) Disposition of toxic drugs and chemicals in man, 9th edn. Biomedical Publications, Seal Beach, CA

2. Goodman LS, Brunton LL, Chabner B, Knollmann BC (2011) Goodman & Gilman's pharmacological basis of therapeutics, 12th edn. McGraw-Hill, New York

3. Gradinaru J, Vullioud A, Eap CB, Ansermot N (2014) Quantification of typical antipsychotics in human plasma by ultra-high performance liquid chromatography tandem mass spectrometry for therapeutic drug monitoring. J Pharm Biomed Anal 88:36–44. doi:10.1016/j.jpba.2013.07.041

4. Juenke JM, Brown PI, Urry FM, Johnson-Davis KL, McMillin GA (2013) Simultaneous UPLC-MS/MS assay for the detection of the traditional antipsychotics haloperidol, fluphenazine, perphenazine, and thiothixene in serum and plasma. Clin Chim Acta 423:32–34. doi:10.1016/j.cca.2013.04.014

# Chapter 7

# Quantitation of Total Buprenorphine and Norbuprenorphine in Meconium by LC-MS/MS

## Stephanie J. Marin and Gwendolyn A. McMillin

## Abstract

Buprenorphine (Suboxone, Zubsolv, Buprenex, Butrans, etc.) is an opioid drug that has been used to treat opioid dependence on an outpatient basis, and is also prescribed for managing moderate to severe pain. Pregnant women may be prescribed buprenorphine as part of a treatment plan for opioid addiction. This chapter quantitates buprenorphine and norbuprenorphine in meconium by liquid chromatography tandem mass spectrometry (LC-MS/MS).

**Key words** Buprenorphine, LC-MS/MS, Meconium, Neonate, Neonatal abstinence syndrome, NAS, Pregnancy

## 1 Introduction

Buprenorphine was approved for treatment of opioid addiction in 2002. Studies have shown it to be a safe and effective treatment for pregnant women [1]. Babies born to mothers prescribed buprenorphine instead of methadone required less treatment for Neonatal Abstinence Syndrome (NAS). Kacinko et al. demonstrated a correlation between meconium buprenorphine concentrations and neonatal outcomes [2]. Buprenorphine concentrations in meconium could be used to guide newborn treatment and verify the mother's compliance with prescribed therapy. The major metabolite of buprenorphine is norbuprenorphine. Both compounds are pharmacologically active and both are extensively metabolized to their glucuronide conjugates, which are also active metabolites. This chapter quantitates total buprenorphine and norbuprenorphine in meconium by liquid chromatography tandem mass spectrometry (LC-MS/MS). Specimens undergo pre-analytical enzyme hydrolysis to convert the glucuronides to free drug. Sample cleanup using solid phase extraction (SPE) is conducted prior to analysis by LC-MS/MS.

Uttam Garg (ed.), *Clinical Applications of Mass Spectrometry in Drug Analysis: Methods and Protocols*, Methods in Molecular Biology, vol. 1383, DOI 10.1007/978-1-4939-3252-8_7, © Springer Science+Business Media New York 2016

## 2  Materials

### 2.1  Samples

Meconium (blackish material), the first stool of the newborn, typically passed during the first 48 h after birth.

### 2.2  Reagents

1. Drug-free meconium: residual meconium specimens that did not contain buprenorphine analytes pooled and stored refrigerated (2–8 °C).

2. 0.1 M sodium acetate buffer, pH 5: Stable at room or refrigerated (2–8 °C) for at least 7 days.

3. Beta-glucuronidase enzyme with purity determined and a certificate of analysis (typically 100,000 units/mL, Campbell Science, Rockford, IL). This is used to prepare the beta-glucuronidase solution. Stable refrigerated until expiration date on bottle.

4. Beta-glucuronidase enzyme solution, 5000 units/mL, prepared in pH 5 sodium acetate buffer. Stable at refrigerated (2–8 °C) for 3 months.

5. Ammonium hydroxide: 28–30 % $NH_3$.

6. Elution solvent:ethyl acetate:2-propanol:ammonium hydroxide (75:20:5). The elution solvent must be freshly prepared just prior to its use in each batch run.

7. Ultrapure water with a resistance of 18 M$\Omega$ or greater (e.g. Nanopure).

8. Acetonitrile: HPLC grade.

9. Mobile phase A (0.1 % formic acid in water): Add 2 mL of 98 % formic acid to 2 L of ultrapure water. Stable for 10 days at room temperature.

10. Mobile phase B (0.1 % formic acid in acetonitrile): Add 2 mL of 98 % formic acid to 2 L of acetonitrile. Stable for 10 days at room temperature.

11. Autosampler wash 1 (Mobile Phase A: Acetonitrile, 95:5): Add 100 mL of acetonitrile to 1900 mL of Mobile Phase A. Stable for 14 days at room temperature (see **Note 1**).

12. Autosampler wash 2 (2-propanol:acetonitrile:methanol, 60:20:20): Prepare by mixing 1200 mL 2-propanol, 400 mL acetonitrile, and 400 mL of methanol in a 2 L bottle. Stable for 14 days at room temperature (see **Note 1**).

### 2.3  Standards and Calibrators

1. Buprenorphine: 1.0 mg/mL in methanol (Cerilliant, Round Rock, TX).

2. Norbuprenorphine: 1.0 mg/mL in methanol (Cerilliant, Round Rock, TX).

3. Buprenorphine and Norbuprenorphine Calibrator Spike solution (0.625 ng/µL in methanol): prepared by adding 62.5 µL

**Table 1**
**Concentration of calibrators (Cal 1–4)**

| Analyte | Final conc. (ng/g) |
|---------|--------------------|
| Buprenorphine<br>Norbuprenorphine | Cal 1 = 20<br>Cal 2 = 250<br>Cal 3 = 750<br>Cal 4 = 1000 |

**Table 2**
**Concentration of controls**

| Analyte | Final conc. (ng/g) |
|---------|--------------------|
| Buprenorphine[a]<br>Norbuprenorphine[a] | Low = 40<br>High = 800 |

[a]Prepared as glucuronide, calculated as free drug and metabolite

of 1.0 mg/mL buprenorphine and 62.5 μL of 1.0 mg/mL norbuprenorphine to 100 mL of methanol in a Class A volumetric flask (*see* **Note 2** and Table 1).

*2.4 Quality Control and Internal Standard*

1. Buprenorphine glucuronide: 0.1 mg/mL in methanol (Cerilliant, Round Rock, TX).

2. Norbuprenorphine glucuronide: 0.1 mg/mL in methanol (Cerilliant, Round Rock, TX).

3. Buprenorphine and Norbuprenorphine Glucuronide Control Spike Solution (0.710 ng/μL in methanol): prepared by adding 710 μL of 0.1 mg/mL buprenorphine glucuronide and 710 μL of norbuprenorphine glucuronide to 100 mL of methanol in a Class A volumetric flask (*see* **Note 3** and Table 2).

4. Buprenorphine-D$_4$: 1.0 mg/mL in methanol (Cerilliant, Round Rock, TX).

5. Norbuprenorphine-D$_3$: 1.0 mg/mL in methanol (Cerilliant, Round Rock, TX).

6. Internal Standard Spike Solution (5 ng/μL in methanol). Prepared by adding 500 μL of buprenorphine-D$_4$ and 500 μL of norbuprenorphine-D$_3$ to 100 mL of methanol in a Class A volumetric flask.

*2.5 Supplies*

1. 5 mL polypropylene conical centrifuge tube with snap cap.

2. Applicator, Plain Wood, 6″.

3. Strata XL-C SPE column, 60 mg/3 mL (Phenomenex, Torrance, CA) (*see* **Note 4**).

4. Vial, Clear, Silanized, Snap-It, 12 × 32 mm, 1.5 mL.

5. Cap, Snap w/preslit, Blue, PTFE/White Silicone.

6. Agilent Poroshell 120 EC-C$_{18}$, 2.7 μm, 50×3 mm (Agilent Technologies, Santa Clara, CA).

7. Stainless steel beads for Bullet Blender™, 0.9–2.0 mm blend stainless steel beads, #SSB14B (Next Advance, Inc., Averill Park, NY).

8. Standard glassware, pipette tips, and other supplies.

*2.6  Equipment*

1. Bullet Blender™ (Next Advance, Inc., Averill, NY).

2. 48-place positive pressure manifold (SPEWare, Baldwin Park, CA or Biotage Charlotte, NC) (*see* **Note 5**).

3. 48-place sample concentrator: Cerex (SPEWare, Baldwin Park, CA) or Turbovap (Biotage, Charlotte, NC).

4. Heat block.

5. AB SCIEX Triple Quad™ 5500 mass spectrometer interfaced with CTC PAL HTC-*xt*-DLW autosampler and Agilent 1260 Infinity series binary pump, degasser, and column oven, operated in positive electrospray ionization mode. Instrument control is performed using AB SCIEX Analyst® software (*see* **Note 6**).

   (a)  LC Conditions:

   - Flow rate: 0.50 mL/min.
   - Column Temp: 35 °C.
   - LC Gradient (*see* Table 3).

   (b)  Autosampler settings (*see* Table 4).

   (c)  Ion source settings (*see* Table 5).

   (d)  MRM transitions and associated MS parameters (*see* Table 6).

**Table 3**
**LC gradient conditions**

| Time | %B |
| --- | --- |
| 0.00 | 25 |
| 2.00 | 25 |
| 3.50 | 95 |
| 4.00 | 95 |
| 4.10 | 25 |
| 5.00 | 25 |

**Table 4**
**Autosampler parameters**

| Parameter | Value |
| --- | --- |
| Cycle | Analyst LC injDLW Standard_Rev05 |
| Syringe | 100 μL DLW |
| Loop 1 volume (mL) | 5 |
| Loop 2 volume (μL) | 100 |
| Actual syringe (μL) | 100 |
| Injection volume (μL) | 5 |
| Airgap volume (μL) | 3 |
| Front volume (μL) | 3 |
| Rear volume (μL) | 3 |
| Filling speed (μL/s) | 5 |
| Pump delay 3 ms | 3 |
| Inject to: LCvlv1 | LCvlv1 |
| Injection speed (μL/s) | 5 |
| Pre inject delay (ms) | 500 |
| Post inject delay (ms) | 500 |
| Needle gap (μm) | 3 |
| Valve clean time solvent 2 (s) | 6 |
| Post clean time solvent 2 (s) | 6 |
| Valve clean time solvent 1 (s) | 6 |
| Post clean time solvent 1 (s) | 6 |
| Stator wash | 1 |
| Delay stator wash (s) | 120 |
| Stator wash time (s) | 5 |
| Stator wash time solvent 2 (s) | 5 |
| Stator wash time solvent 1 (s) | 5 |

# 3    Methods

### 3.1    Procedure

1. Accurately weigh $0.25 \pm 0.05$ g of drug free meconium to tubes 1–7 using a plain wood applicator.

2. To each test specimen tube, accurately weigh $0.25 \pm 0.05$ g of meconium specimen using a new, clean plain wood applicator

**Table 5**
**MS source parameters**

| Parameter | Value |
|---|---|
| CAD (Collision gas, psi) | 8 |
| CUR (Curtain gas, psi) | 30 |
| GS1 (Nebulizer gas, psi) | 30 |
| GS2 (Heater gas, psi) | 50 |
| TEM (Temperature, °C) | 600 |
| IS (IonSpray voltage, V) | 4000 |
| Entrance potential (mV) | 10 |
| Dwell time (ms) | 10 |
| Declustering potential (V) | 50 |
| Resolution | unit |

**Table 6**
**MRM transitions and MS parameters**

| Compound | Q1 | Q3 | Collision energy (V) | Exit potential (V) |
|---|---|---|---|---|
| Buprenorphine D4 1 | 472.3 | 415.1 | 47 | 20 |
| Buprenorphine D4 2 | 472.3 | 400.0 | 52 | 20 |
| Norbuprenorphine D3 1 | 417.3 | 343.2 | 40 | 15 |
| Norbuprenorphine D3 2 | 417.3 | 187.0 | 50 | 10 |
| Buprenorphine 1 | 468.2 | 414.4 | 47 | 18 |
| Buprenorphine 2 | 468.2 | 396.1 | 53 | 18 |
| Norbuprenorphine 1 | 414.2 | 187.1 | 49 | 9 |
| Norbuprenorphine 2 | 414.2 | 211.2 | 49 | 10 |

for each specimen. Keep the walls of the tubes as clean as possible and the meconium as close to the bottom as possible. To conserve specimen, a twofold dilution may be performed (*see* **Note 7**).

3. Spike the calibrators:

   (a) Add 8 μL of the Buprenorphine and Norbuprenorphine Calibrator Spike Solution to tube (calibrator) 1.

   (b) Add 100 μL of the Buprenorphine and Norbuprenorphine Calibrator Spike Solution to tube (calibrator) 2.

(c)  Add 300 µL of the Buprenorphine and Norbuprenorphine Calibrator Spike solution to tube (calibrator) 3.

(d)  Add 400 µL of the Buprenorphine and Norbuprenorphine Calibrator Spike solution to tube (calibrator) 4.

4. Spike the controls.

(a)  Add 20 µL of the Buprenorphine and Norbuprenorphine Glucuronide Control Spike Solution to tube 5 (low positive control).

(b)  Add 400 µL of the Buprenorphine and Norbuprenorphine Glucuronide Control Spike Solution to tube 6 (high positive control).

(c)  Tube 7 is a negative control that contains only drug-free meconium.

5. Using a repeating pipette, add 10 µL of the Internal Standard Spike Solution to each tube.

6. Add 1.0 mL of beta-glucuronidase enzyme solution to each tube.

7. Add 1/8 teaspoon (0.6 mL) of 0.9–2.0 mm stainless steel beads to each tube.

8. Cap the tubes securely and place in the Bullet Blender™ at setting 6 for approximately 2 min.

9. Verify that all specimens have a uniform appearance and no large meconium pieces are apparent; a smooth semi-liquid should be obtained.

10. Hydrolyze the homogenized samples in the heat block at 70 °C for 1 h.

11. Cool briefly and add 1.5 mL pH 5 sodium acetate buffer.

12. Centrifuge each sample for 15 min at 0 °C and $23{,}447 \times g$. The low temperature congeals most of the soluble lipids and the high speed centrifugation is necessary to obtain the clearest supernatant possible.

13. Condition the Strata XL-C SPE columns with 1 mL of methanol. Elute by gravity.

14. Equilibrate the Strata XL-C SPE columns with 1 mL 0.1 M pH 5 sodium acetate buffer. Elute by gravity.

15. Load samples in proper order onto the columns. Adjust the pressure to one drop per 4 s.

16. Wash columns at one drop per second with 1 mL of 0.1 M pH 5 sodium acetate buffer.

17. Wash columns at one drop per second with 1 mL of methanol.

18. Dry the columns for 4 min at 25 psi.

19. Elute BY GRAVITY with $2 \times 0.60$ mL of freshly prepared elution solvent (75:20:5 ethyl acetate:2-propanol:ammonium hydroxide) into corresponding pre-labeled 1.5 mL vials.

20. Place the autosampler vials in the Turbovap at 40 °C and 3–5 psi for approximately 20 min, OR place in the SPEWare 48-place sample concentrator for 25 min at 40 °C and 25 psi. Evaporate to dryness.

21. Reconstiue the dried extracts in 100 µL of 90:10 water:acetonitrile. Cap and briefly vortex.

22. Perform instrumental analysis.

**3.2 Data Analysis**

Data analysis and quantitation is performed using AB SCIEX MultiQuant™ software. LOD, LLOQ, and ULOQ for all analytes are listed in Table 7 (*see* **Note 8**).

Typical extracted ion chromatograms for a calibrator and a patient sample are shown in Figs. 1 and 2.

**Table 7**
**Reportable range of results in ng/g**

| Analyte | LOD | LLOQ | ULOQ |
|---|---|---|---|
| Buprenorphine | 10 | 20 | 1000 |
| Norbuprenorphine | 10 | 20 | 1000 |

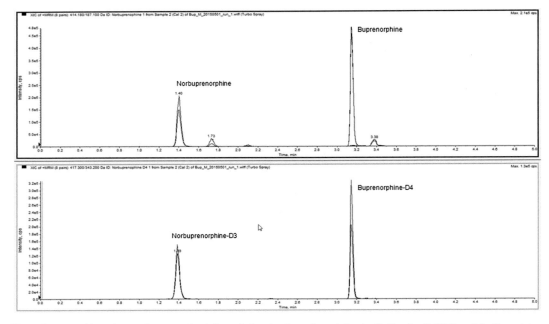

**Fig. 1** Extracted ion chromatograms for internal standards and analytes in Calibrator 2 (250 ng/g) all analytes, 200 ng/g for each internal standard

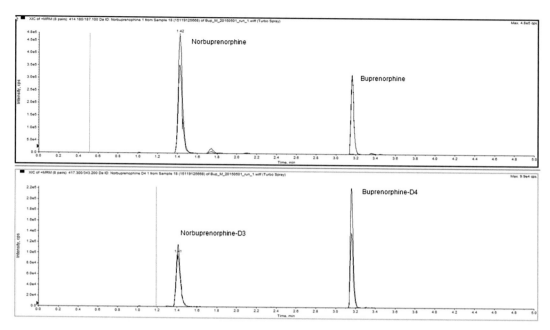

**Fig. 2** Extracted ion chromatograms for internal standards and analytes in a positive patient specimen. Buprenorphine: 247 ng/g, norbuprenorphine: 740 ng/g

---

## 4   Notes

1. Several autosampler wash solvents were evaluated. Analytical carryover of buprenorphine was observed with most wash solvents. The stronger autosampler wash 2 is critical for eliminating carryover. Potential for carryover is dependent on the autosampler and should be re-evaluated if a different autosampler is used.

2. Calibrators are prepared by spiking an aliquot of drug-free meconium with a calibrator spike solution prepared in methanol.

3. Calibrators are prepared using buprenorphine and norbuprenorphine; controls are prepared from a spike solution prepared with buprenorphine glucuronide and norbuprenorphine glucuronide. The positive controls therefore serve as a both analyte controls and hydrolysis controls. The controls are prepared to produce free buprenorphine and norbuprenorphine at 40 ng/g (2× the cutoff) and 800 ng/g (80 % of ULOQ) after hydrolysis.

4. The use of the Strata XL-C and not Strata X-C SPE columns is critical for assay performance. Column plugging was an issue when the Strata X-C columns were used. This was eliminated by using the Strata XL-C columns.

5. This method was validated using positive pressure for SPE extraction. The use of vacuum manifolds was not evaluated.

6. LC and MS parameters will vary from instrument to instrument. The parameters shown here can be used as a starting point, but MS parameters for all compounds should be optimized for the instrument used. If a different LC is used, chromatographic conditions may need to be modified.

7. A half-weight (2× dilution) can be made if there is a need to conserve specimen or there is insufficient specimen to perform the analysis. Accurately weigh $0.125 \pm 0.05$ g of meconium. Add the same amount of the Internal Standard Spike Solution. Adjust calculations and cutoff concentrations accordingly.

8. Calibration curves were a linear fit, weighted $1/x$. The MQ4 integrator was used for peak integration. Integration parameters will need to be determined based on the quality of the raw data to provide the most consistent and accurate peak integration. Acceptance criteria determined during validation include: each calibrator was accurate to $\pm 15$ % of its target concentration and the correlation coefficient of the linear fit was >0.995. Controls were $\pm 20$ of the target value. Ion mass ratios were within $\pm 25$ %.

## References

1. Jones HE (2010) Neonatal abstinence syndrome after methadone or buprenorphine exposure. N Engl J Med 363(24):2320–2331

2. Kacinko SL et al (2008) Correlations of Maternal Buprenorphine Dose, Buprenorphine, and Metabolite Concentrations in Meconium with Neonatal Outcomes. Clin Pharmacol Ther 84(5):604–612

# Chapter 8

## Quantitation of Buprenorphine, Norbuprenorphine, Buprenorphine Glucuronide, Norbuprenorphine Glucuronide, and Naloxone in Urine by LC-MS/MS

### Stephanie J. Marin and Gwendolyn A. McMillin

### Abstract

Buprenorphine is an opioid drug that has been used to treat opioid dependence on an outpatient basis, and is also prescribed for managing moderate to severe pain. Some formulations of buprenorphine also contain naloxone to discourage misuse. The major metabolite of buprenorphine is norbuprenorphine. Both compounds are pharmacologically active and both are extensively metabolized to their glucuronide conjugates, which are also active metabolites. Direct quantitation of the glucuronide conjugates in conjunction with free buprenorphine, norbuprenorphine, and naloxone in urine can distinguish compliance with prescribed therapy from specimen adulteration intended to mimic compliance with prescribed buprenorphine.

This chapter quantitates buprenorphine, norbuprenorphine, their glucuronide conjugates and naloxone directly in urine by liquid chromatography tandem mass spectrometry (LC-MS/MS). Urine is pretreated with formic acid and undergoes solid phase extraction (SPE) prior to analysis by LC-MS/MS.

**Key words** Buprenorphine, LC/MS/MS, Urine, Suboxone, Zubsolv

## 1   Introduction

Opioid drugs are among the most powerful physically addicting drugs known. Physical dependence and tolerance can develop rapidly, and abuse is widespread. Several buprenorphine formulations (e.g. Subutex, Buprenex, Butrans) have been approved by the US Food and Drug Administration for use in treating opioid addiction and moderate-to-severe pain. Naloxone is an opioid antagonist added to some buprenorphine formulations (Suboxone, Zubsolv) to prevent misuse [1]. When oral formulations containing naloxone are taken correctly, the bioavailability of naloxone is minimal, such that naloxone will not interfere with the pharmacological activity of buprenorphine. When such formulations are misused, such as through administration by insufflation (snorting) or injection, the naloxone bioavailability is high, and naloxone will

Uttam Garg (ed.), *Clinical Applications of Mass Spectrometry in Drug Analysis: Methods and Protocols*, Methods in Molecular Biology, vol. 1383, DOI 10.1007/978-1-4939-3252-8_8, © Springer Science+Business Media New York 2016

compete with buprenorphine for opioid activity, in some cases precipitating opioid withdrawal symptoms. Urine is a common specimen used to evaluate compliance with prescribed buprenorphine due to its ease of collection and likelihood of detection compared to blood. However, urine collections are not routinely observed in a medical setting and may be subject to adulteration by the donor to mimic compliance. The presence of free naloxone in a urine specimen may suggest adulteration of the specimen by adding drug directly into the urine [2, 3].

Opioids are commonly determined by methods that involve hydrolyzing the glucuronide metabolites to the free form and measuring the total concentration of the free and glucuronide analytes combined. This practice increases the concentration of free drug and the likelihood of detection. However, in this method the glucuronides of buprenorphine and norbuprenorphine are measured directly [2, 3]. The pattern of glucuronide metabolites may help define the metabolic phenotype for an individual patient, and may suggest time of last use. For example, norbuprenorphine glucuronide is likely to persist in urine the longest of the analytes included in this assay, and may suggest that time of last dose was not recent if this is the only analyte detected. Quantitative results are reported for all analytes.

## 2    Materials

### 2.1    Samples

Patient random urine specimens are collected using a standard urine cup with no preservative.

### 2.2    Reagents

1. Drug-free urine: collected with no preservative and qualified to assure that no buprenorphine or naloxone analytes are present.

2. Ammonium hydroxide: 28–30 % $NH_3$.

3. Ultrapure water with a resistance of 18 $M\Omega$ or greater (e.g. Nanopure).

4. 0.1 M formic acid: Add 4 mL 98–100 % formic acid to 1 L of ultrapure water. Stable for 7 days at room temperature.

5. Mobile phase A (0.1 % formic acid in water): Add 2 mL of 98 % formic acid to 2 L of ultrapure water. Stable for 10 days at room temperature.

6. Mobile phase B (0.1 % formic acid in acetonitrile): Add 2 mL of 98 % formic acid to 2 L of acetonitrile. Stable for 10 days at room temperature.

7. Autosampler wash 1 (Mobile Phase A: Acetonitrile, 95:5): Add 100 mL of acetonitrile to 1900 mL of Mobile Phase A. Stable for 14 days at room temperature (*see* **Note 1**).

8. Autosampler wash 2 (2-propanol:acetonitrile:methanol, 60:20:20): Prepare by mixing 1200 mL 2-propanol, 400 mL acetonitrile, and 400 mL of methanol in a 2 L bottle. Stable for 14 days at room temperature (*see* **Note 1**).

**2.3 Standards and Calibrators**

1. Buprenorphine: 1.0 mg/mL in methanol (Cerilliant, Round Rock, TX).

2. Norbuprenorphine: 1.0 mg/mL in methanol (Cerilliant, Round Rock, TX).

3. Buprenorphine glucuronide: 0.1 mg/mL in methanol (Cerilliant, Round Rock, TX).

4. Norbuprenorphine glucuronide: 0.1 mg/mL in methanol (Cerilliant, Round Rock, TX).

5. Naloxone: 1.0 mg/mL in methanol (Cerilliant, Round Rock, TX).

6. Buprenorphine and norbuprenorphine: 10 ng/μL in methanol. Prepared by adding 100 μL each 1.0 mg/mL buprenorphine and 1.0 mg/mL norbuprenorphine to the same 10 mL Class A volumetric flask. Dilute to the mark with methanol.

7. Calibrators prepared in drug-free urine (*see* Table 1). Prepared by adding the amount listed in Table 2 for each calibrator to 100 mL of drug-free urine in a Class A volumetric flask. Store at −80 °C.

**2.4 Quality Control and Internal Standard**

1. Buprenorphine-$D_4$: 1.0 mg/mL in methanol.

2. Norbuprenorphine-$D_3$: 1.0 mg/mL in methanol.

3. Naloxone-$D_5$: 1.0 mg/mL in methanol.

4. Buprenorphine low and high positive controls prepared in drug-free urine (*see* Table 3). Prepared by adding the amount listed in Table 4 for each control to 100 mL of drug-free urine in a Class A volumetric flask. Store at −80 °C.

5. Buprenorphine internal standard (IS) master mix spike solution: 4 ng/μL each buprenorphine-$D_4$, norbuprenorphine-$D_3$, naloxone-$D_5$ prepared in methanol. Prepared by adding 400 μL of buprenorphine-$D_4$, norbuprenorphine-$D_3$, naloxone-$D_5$ (all 1.0 mg/mL) to methanol in a 100 mL Class A volumetric flask. Store at −80 °C.

6. Buprenorphine IS master mix (40 ng/mL): Prepared by adding 0.5 mL buprenorphine IS master mix spike solution to 50 mL 0.1 M formic acid. Prepared fresh before each sample extraction.

**2.5 Supplies**

1. Solid phase extraction columns: PSCX, 10 mg/1 mL (SPEware Inc., San Pedro, CA) or Strata X-C, 10 mg/1 mL (Phenomenex, Torrance CA), single columns or 96-well plate (*see* **Note 2**).

**Table 1**
**Concentration of calibrators (Cal 1–6)**

| Analyte | Final conc. (ng/mL) |
|---|---|
| Buprenorphine<br><br>Norbuprenorphine | Cal 1 = 2<br>Cal 2 = 30<br>Cal 3 = 200<br>Cal 4 = 400<br>Cal 5 = 800<br>Cal 6 = 1000 |
| Buprenorphine glucuronide<br><br>Norbuprenorphine glucuronide | Cal 1 = 5<br>Cal 2 = 30<br>Cal 3 = 200<br>Cal 4 = 400<br>Cal 5 = 800<br>Cal 6 = 1000 |
| Naloxone | Cal 1 = 100<br>Cal 2 = 150<br>Cal 3 = 200<br>Cal 4 = 400<br>Cal 5 = 800<br>Cal 6 = 1000 |

**Table 2**
**Preparation of calibrators**

| Calibrator | Buprenorphine and norbuprenorphine 10 ng/ μL | Buprenorphine 1.0 mg/mL | Norbuprenorphine 1.0 mg/mL | Buprenorphine glucuronide 0.1 mg/mL | Norbuprenorphine glucuronide 0.1 mg/mL | Naloxone 1.0 mg/mL |
|---|---|---|---|---|---|---|
| Cal 1 | 20 μL | – | – | 5.0 μL | 5.0 μL | 10 μL |
| Cal 2 | 300 μL | – | – | 30 μL | 30 μL | 15 μL |
| Cal 3 | – | 20 μL | 20 μL | 200 μL | 200 μL | 20 μL |
| Cal 4 | – | 40 μL | 40 μL | 400 μL | 400 μL | 40 μL |
| Cal 5 | – | 80 μL | 80 μL | 800 μL | 800 μL | 80 μL |
| Cal 6 | – | 100 μL | 100 μL | 1000 μL | 1000 μL | 100 μL |

**Table 3**
**Concentration of controls**

| Analyte | Final conc. (ng/mL) |
|---|---|
| Buprenorphine | Low = 4 ng/mL |
| Norbuprenorphine | High = 800 ng/mL |
| Buprenorphine glucuronide | Low = 10 ng/mL |
| Norbuprenorphine glucuronide | High = 800 ng/mL |
| Naloxone | Low = 200 ng/mL<br>High = 800 ng/mL |

**Table 4**
**Preparation of Controls**

| Control | Buprenorphine and norbuprenorphine 10 ng/μL | Buprenorphine 1.0 mg/mL | Norbuprenorphine 1.0 mg/mL | Buprenorphine glucuronide 0.1 mg/mL | Norbuprenorphine glucuronide 0.1 mg/mL | Naloxone 1.0 mg/mL |
|---|---|---|---|---|---|---|
| Low positive control | 40 μL | – | – | 10 μL | 10 μL | 20 μL |
| High positive control | – | 80 μL | 80 μL | 800 μL | 800 μL | 80 μL |

2. 96-well polypropylene transfer plates.

3. 96-well polypropylene collection plates.

4. Silicone plate cover.

5. LC column: Agilent Poroshell 120 EC-C18, 2.7 μm, 50 × 3 mm (Agilent Technologies, Santa Clara, CA).

*2.6 Equipment*

1. Liquid handler (unless pipetting manually): JANUS model AJM8001 or comparable instrument (Perkin-Elmer, Waltham, Massachusetts) (*see* **Note 3**).

2. 96-place positive pressure automated manifold: Cerex (Speware, Baldwin Park, CA) or comparable instrument (*see* **Note 4**).

3. 96-place sample concentrator: Cerex (Speware, Baldwin Park, CA) or comparable instrument.

4. AB SCIEX Triple Quad™ 5500 mass spectrometer interfaced with CTC PAL HTC-*xt*-DLW autosampler and Agilent 1260 Infinity series binary pump, degasser, and column oven, operated in positive electrospray ionization mode. Instrument control is performed using AB SCIEX Analyst® software (*see* **Note 5**).

(a) LC Conditions:

- Flow rate: 0.50 mL/min.
- Column Temp: 35 °C.
- LC Gradient (*see* Table 5).

(b) Diverter valve: flow directed into mass spectrometer from 1.0 to 2.5 min of the chromatographic run and diverted to waste outside of that time interval.

(c) Autosampler settings (*see* Table 6).

(d) Ion source settings (*see* Table 7).

(e) MRM transitions and associated MS parameters (*see* Table 8).

**Table 5**
**LC gradient conditions**

| Time | %B |
|------|-----|
| 0.00 | 5 |
| 0.25 | 5 |
| 2.00 | 75 |
| 2.50 | 95 |
| 3.00 | 95 |
| 3.10 | 5 |
| 5.00 | 5 |

## 3   Methods

### 3.1   Procedure

1. Condition SPE columns with 0.5 mL methanol followed by 0.5 mL 0.1 M formic acid.
2. Organize calibrators, controls, and specimens.
3. Add 0.25 mL of specimen, calibrator, or QC into the appropriate well of a 96-well polypropylene transfer plate.
4. Add 0.25 mL of buprenorphine IS master mix to each sample.
5. Mix and transfer to the conditioned SPE columns.
6. Load the samples onto the columns at one drop/s.
7. Wash each column with 0.5 mL of 0.1 M formic acid.
8. Wash each column with 1 mL of methanol.
9. Dry the columns under nitrogen for 1 min at 40 psi and 40 °C.
10. Elute the analytes with two 0.5 mL aliquots of 68:30:2 ethyl acetate:methanol:ammonium hydroxide into the 2 mL collection plate.
11. Dry under nitrogen at 40 °C.
12. Reconstitute in 300 μL of 90:10 water:acetonitrile. Seal with a plate cover and vortex for 15 s.

### 3.2   Data Analysis

Data analysis and quantitation is performed using AB SCIEX MultiQuant™ software. LOD, LLOQ, and ULOQ for all analytes are listed in Table 9 (*see* **Note 6**). Typical extracted ion chromatograms for a calibrator and a patient sample are shown in Figs. 1 and 2.

**Table 6**
**Autosampler parameters**

| Parameter | Value |
|---|---|
| Cycle | Analyst LC injDL W Stadard_Rev05 |
| Syringe | 100 μL DLW |
| Loop 1 volume (mL) | 5 |
| Loop 2 volume (μL) | 100 |
| Actual syringe (μL) | 100 |
| Injection volume (μL) | 5 |
| Airgap volume (μL) | 3 |
| Front volume (μL) | 3 |
| Rear volume (μL) | 3 |
| Filling speed (μL/s) | 5 |
| Pump delay 3 ms | 3 |
| Inject to: LCvlv1 | LCvlv1 |
| Injection speed (μL/s) | 5 |
| Pre inject delay (ms) | 500 |
| Post inject delay (ms) | 500 |
| Needle gap (μm) | 3 |
| Valve clean time solvent 2 (s) | 6 |
| Post clean time solvent 2 (s) | 6 |
| Valve clean time solvent 1 (s) | 6 |
| Post clean time solvent 1 (s) | 6 |
| Stator wash | 1 |
| Delay stator wash (s) | 120 |
| Stator wash time (s) | 5 |
| Stator wash time solvent 2 (s) | 5 |
| Stator wash time solvent 1 (s) | 5 |

## 4    Notes

1. Several autosampler wash solvents were evaluated and carry-over for buprenorphine was observed if these wash solvents were not used. The stronger autosampler wash 2 is critical for eliminating carryover. This should be re-evaluated if a different autosampler is used.

**Table 7**
**MS source parameters**

| Parameter | Value |
|---|---|
| CAD (Collision gas, psi) | 10 |
| CUR (Curtain gas, psi) | 10 |
| GS1 (Nebulizer gas, psi) | 50 |
| GS2 (Heater gas, psi) | 80 |
| TEM (Temperature °C) | 600 |
| IS (IonSpray voltage, V) | 5000 |
| Entrance potential (mV) | 10 |
| Entrance potential (mV) | 10 |
| Dwell time (ms) | 10 |
| Resolution | UNIT |

**Table 8**
**MRM transitions and MS parameters**

| Group | Compound | Q1 Mass (Da) | Q3 Mass (Da) | Dwell (ms) | DP | CE | CXP |
|---|---|---|---|---|---|---|---|
| Buprenorphine | Buprenorphine D4 1 | 472.30 | 415.10 | 10 | 50 | 47 | 20 |
| | Buprenorphine D4 2 | 472.30 | 400.00 | 10 | 50 | 52 | 20 |
| | Buprenorphine 1 | 468.24 | 396.24 | 10 | 300 | 53 | 18 |
| | Buprenorphine 2 | 468.24 | 414.29 | 10 | 300 | 47 | 18 |
| | Buprenorphine gluc 1 | 644.42 | 396.10 | 10 | 110 | 47 | 18 |
| | Buprenorphine gluc 2 | 644.42 | 414.30 | 10 | 110 | 47 | 18 |
| Norbuprenorphine | Norbuprenorphine D3 1 | 417.30 | 343.20 | 10 | 50 | 40 | 15 |
| | Norbuprenorphine D3 2 | 417.30 | 187.00 | 10 | 50 | 50 | 10 |
| | Norbuprenorphine 1 | 414.18 | 187.10 | 10 | 250 | 49 | 8 |
| | Norbuprenorphine 2 | 414.18 | 211.20 | 10 | 250 | 49 | 10 |
| | Norbuprenorphine gluc 1 | 590.19 | 396.30 | 10 | 250 | 51 | 18 |
| | Norbuprenorphine gluc 2 | 590.19 | 165.00 | 10 | 250 | 51 | 18 |
| Naloxone | Naloxone D5 1 | 333.20 | 258.10 | 10 | 150 | 35 | 15 |
| | Naloxone D5 2 | 333.20 | 273.00 | 10 | 150 | 35 | 15 |
| | Naloxone 1 | 328.11 | 183.10 | 10 | 130 | 47 | 7 |
| | Naloxone 2 | 328.11 | 127.00 | 10 | 130 | 47 | 7 |

*DP* declustering potential, *CE* collision energy, *CXP* exit potential

**Table 9**
**Reportable range of analytes**

| Analyte | LOD | LLOQ | ULOQ |
| --- | --- | --- | --- |
| Buprenorphine | 1 | 2 | 1000 |
| Norbuprenorphine | 1 | 2 | 1000 |
| Buprenorphine glucuronide | 2.5 | 5 | 1000 |
| Norbuprenorphine glucuronide | 2.5 | 5 | 1000 |
| Naloxone | 50 | 100 | 1000 |

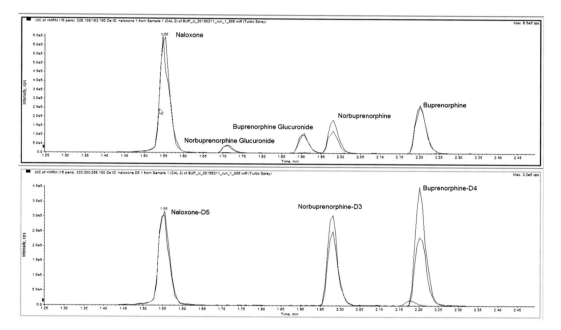

**Fig. 1** Extracted ion chromatograms for internal standards and analytes in Calibrator 2 (30 ng/mL) all analytes, 10 ng/mL for each internal standard

2. The method has been validated using both of these SPE columns with similar results.

3. A JANUS model AJM8001 was used to prepare the samples for analysis, but a comparable liquid handler or manual pipetting can be used.

4. This method was validated using positive pressure for SPE extraction. The use of vacuum manifolds was not evaluated.

5. LC and MS parameters will vary from instrument to instrument. The parameters shown here can be used as a starting point, but MS parameters for all compounds should be opti-

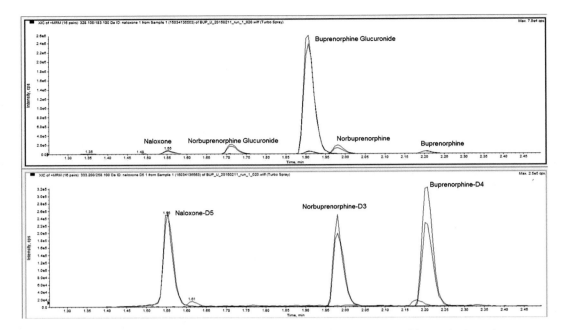

**Fig. 2** Extracted ion chromatograms for internal standards and analytes in a positive patient specimen

mized for the instrument used. If a different LC is used, chromatographic conditions may need to be modified.

6. Calibration curves were a linear fit, weighted 1/x, and forced through zero. The MQ4 integrator was used for peak integration. Integration parameters will need to be determined based on the quality of the raw data to provide the most consistent and accurate peak integration. Acceptance criteria determined during validation include: each calibrator was accurate to ±15 % of its target concentration and the correlation coefficient of the linear fit was >0.995. Controls were ±20 of the target value. Ion mass ratios were within ±20 %.

# References

1. Heikman P et al (2013) Urine naloxone concentration at different phases of buprenorphine maintenance treatment. Drug Test Anal 6(3): 220–225

2. Al-Asmari AI, Anderson RA (2008) Comparison of nonhydrolysis and hydrolysis methods for the determination of buprenorphine metabolites in urine by liquid chromatography-tandem mass spectrometry. J Anal Toxicol 32(9):744–753

3. McMillin GA et al (2012) Patterns of free (unconjugated) buprenorphine, norbuprenorphine, and their glucuronides in urine using liquid chromatography-tandem mass spectrometry. J Anal Toxicol 36(2):81–87

# Chapter 9

# A Simple Liquid Chromatography Tandem Mass Spectrometry Method for Quantitation of Plasma Busulfan

## Shuang Deng, Michael Kiscoan, Clint Frazee, Susan Abdel-Rahman, Jignesh Dalal, and Uttam Garg

## Abstract

Busulfan is an alkylating agent widely used in the ablation of bone marrow cells before hematopoietic stem cell transplant. Due to large intraindividual and interindividual variations, and narrow therapeutic window, therapeutic drug monitoring of busulfan is warranted. A quick and reliable HPLC-MS/MS method was developed for the assay of plasma busulfan. HPLC involved C18 column, and MS/MS was used in electrospray ionization (ESI) positive mode. Quantitation and identification of busulfan was made using various multiple reactions monitoring (MRMs). Isotopic labeled busulfan-$d_8$ was used as the internal standard. The method is linear from 50 to 2500 ng/mL and has with-in run and between-run imprecision of <10 %.

**Key words** Busulfan, Mass spectrometry, Liquid chromatography, Bone marrow transplant, Leukemia

## 1  Introduction

Busulfan is an anti-leukemic DNA-alkylating agent widely used in combination with cyclophosphamide for myeloablative conditioning regimens prior to hematopoietic stem cell transplantation [1–3]. Busulfan has a narrow therapeutic range with significant toxic side effects at high systemic exposure and risk of incomplete myeloablative and graft rejection at low exposure. Therefore, measurement of busulfan is warranted in busulfan dose adjustment and optimal drug exposure [4].

Various methods including immunoassays, gas chromatography (GC) coupled with electron capture detector or mass spectrometry, liquid chromatography coupled with UV detectors or mass spectrometry or fluorescence detectors have been described [5–16]. Due to better specificity, chromatographic methods are preferred. Since busulfan is not a volatile drug, its measurement by gas chromatography is tedious and time-consuming and requires

Uttam Garg (ed.), *Clinical Applications of Mass Spectrometry in Drug Analysis: Methods and Protocols*, Methods in Molecular Biology, vol. 1383, DOI 10.1007/978-1-4939-3252-8_9, © Springer Science+Business Media New York 2016

sample derivatization and extraction [5, 9, 16]. Liquid chromatography mass spectrometry methods often require sample extraction but are preferred as they do not require sample derivatization. Here, we describe a simple protein precipitation no-extraction LC-MS/MS method for the determination of busulfan. The method uses positive ion electrospray ionization (ESI), multiple reactions monitoring (MRM), and $D_8$-busulfan as internal standard.

## 2 Materials

### 2.1 Samples

1 mL blood in sodium heparin (no gel). Process the sample and analyze within 4 h of collection or freeze plasma at –70 °C until analysis (see **Note 1**). Children receiving busulfan every 6 h with a 120 min infusion have plasma samples drawn at 120, 135, 150, 180, 240, 300, and 360 min from the start of the infusion. Children receiving busulfan every 24 h with a 180 min infusion have plasma samples drawn at 180, 195, 240, 300, 360, and 480 min from the start of the infusion.

### 2.2 Reagents

1. 7.5 M Ammonium acetate (Sigma Chemicals, St. Louis, MO).

2. 0.3 N Zinc Sulfate (Sigma Chemicals, St. Louis, MO).

3. Busulfan Powder (Sigma Chemicals, St. Louis, MO).

4. Busulfan, 1 mg/mL (Cerilliant, Rockwood, CA).

5. Busulfan-$d_8$, 1 mg/mL (Cambridge Isotope Laboratories, Inc.).

6. Mobile phase A (20 mM ammonium acetate/water/0.5 % formic acid): To 1 L of HPLC grade water add 2.7 mL of 7.5 M ammonium acetate and 570 μL of 88 % formic acid. Mix and degas. Stable for 1 month when stored at room temperature.

7. Mobile phase B (20 mM ammonium acetate/methanol/0.5 % formic acid): To 1 L of methanol add 2.7 mL of 7.5 M ammonium acetate and 570 μL of 88 % formic acid. Mix and degas. Stable for 1 month when stored at room temperature.

8. Precipitating reagent: Combine 350 mL methanol, 150 mL 0.3 N Zinc Sulfate Solution, 125 μL of 1 mg/mL busulfan-$d_8$ (primary internal standard).

### 2.3 Calibrators and Controls

1. Primary internal standard, Busulfan-$d_8$, 1 mg/mL in acetone: Dissolve 10 mg in 10 mL acetone. Stable for 1 year when stored at –70 °C.

2. Primary (*1°*) standard, Busulfan 1 mg/mL in acetone: Dissolve 100 mg into a 100 mL volumetric flask and q.s. with acetone, stable for 1 year at –70 °C.

3. Secondary (2°) standard, Busulfan 10 μg/mL in negative plasma: Add 250 μL of primary standard into a 25 mL volumetric flask and q.s. with negative plasma to 25 mL. Stable for 1 year at −70 °C.

4. Tertiary (3°) standard, Busulfan 2500 ng/mL in negative plasma: Add 2.5 mL of secondary standard into a 10 mL volumetric flask and q.s. with negative plasma to 10 mL, stable for 1 year at −70 °C (*see* **Note 2**).

5. Quaternary (4°) standard, Busulfan 1000 ng/mL in negative plasma: Add 1 mL of secondary standard (2°) into a 10 mL volumetric flask and q.s. with negative plasma to 10 mL, stable for 1 year at −70 °C (*see* **Note 3**).

6. Negative plasma matrix preparation: Add 850 mg EDTA trisodium salt hydrate to 500 mL pooled expired plasma from blood bank. The plasma first undergoes three cycles of freeze/thaw cycles. Centrifuge the plasma for 5 min at $4600 \times g$ and filter the supernatant. Stable for 1 year at −70 °C (*see* **Note 4**).

7. Preparation of calibrators: Prepare calibrators in negative plasma as described in Table 1. Stable for 1 year at −70 °C.

8. Preparation of controls: Prepare controls in negative plasma as described in Table 2 (*see* **Note 5**). Stable for 1 year at −70 °C.

*2.4 Analytical Equipment and Operating Conditions*

1. Liquid chromatography system: Prominence UFLC system (Schimadzu Scientific Instruments) or equivalent.

2. Analytic column: Supelcosil LC-18, 5 cm × 4.6 mm, 5 μm (Sigma-Aldrich).

3. LC parameters: Flow rate, 0.9 mL/min. Column temperature, 55 °C. HPLC gradient is shown in Table 3.

**Table 1**
**Preparation of calibrators**

| Calibrator | 2° Standard (mL) | 3° Standard (mL) | 4° Standard (mL) | Negative plasma (mL) | Final concentration (ng/mL) |
|---|---|---|---|---|---|
| 1 | | | 0.5 | 9.5 | 50 |
| 2 | | | 1 | 9.0 | 100 |
| 3 | | 1 | | 9.0 | 250 |
| 4 | 0.5 | | | 9.5 | 500 |
| 5 | 1 | | | 9.0 | 1000 |
| 6 | 2.5 | | | 7.5 | 2500 |

Note: All calibrators are stable for 1 year when stored at −70 °C

**Table 2**
**Preparation of quality controls**

| Quality control | 2° Standard (mL) | Negative plasma (mL) | Final concentration (ng/mL) |
|---|---|---|---|
| 1 | 0.15 | 9.85 | 150 |
| 2 | 0.75 | 9.25 | 750 |
| 3 | 2.00 | 8.00 | 2000 |

**Table 3**
**HPLC gradient**

| Time (min) | Mobile phase B % |
|---|---|
| 2 | 2 |
| 4 | 100 |
| 6 | 100 |
| 6.1 | 2 |
| 8 | 2 |

**Table 4**
**MRMs for busulfan and busulfan-D$_8$**

| Analyte | Q1 | Q3 | Qualifier ion |
|---|---|---|---|
| Busulfan | 264 | 151 | 247 |
| Busulfan-d8 | 272 | 159 | 255 |

4. Mass spectrometry: 4000 Qtrap (AB Sciex) or equivalent. Use electrospray ionization source (ESI) and positive polarity mode to monitor ion pairs in multiple reactions monitoring (MRM) mode. MRMs are given in Table 4. Mass spectrometry settings are given in Table 5. Optimized mass spectrometry parameters are given in Table 6.

# 3   Methods

### 3.1   Stepwise Procedure

1. Pipette 100 μL of well-mixed standards, patient plasma and control to a microcentrifuge tube.

2. Add 100 μL 0.9 % NaCl solution and gently vortex to mix.

**Table 5**
**Mass spectrometry settings**

| | |
|---|---|
| Curtain gas (CUR) | 25 |
| Collision gas (GAD) | Medium |
| Ionspray voltage (IS) | 4000 V |
| Temperature (TEM) | 375 °C |
| Ion source gas 1 (GS 1) | 50 |
| Ion source gas 2 (GS 2) | 60 |
| Interface heater | on |

**Table 6**
**Mass spectrometry optimization for various ions**

| Analyte | Q1 (m/z) | Q3 (m/z) | DP (V) | EP (V) | CXP (V) | CE (eV) |
|---|---|---|---|---|---|---|
| Busulfan 1 | 264.0 | 151.1 | 46 | 10 | 8 | 17 |
| Busulfan 2 | 264.0 | 247.1 | 31 | 10 | 10 | 15 |
| $D_8$-Busulfan 1 | 272.1 | 159.1 | 31 | 10 | 14 | 17 |
| $D_8$-Busulfan 2 | 272.1 | 255.1 | 31 | 10 | 10 | 10 |

3. Add 500 μL precipitating/IS reagent, then immediately cap and vortex twice for total 30 s (2× dilutions are performed for each sample at the same time).

4. Centrifuge tubes for 5 min at $12,000 \times g$.

5. Carefully transfer approximately 100 μL of solution into labeled autosampler vials (*see* **Note 6**).

6. Inject 20 μL into LC/MS/MS for analysis.

*3.2 Data Analysis*

1. Data are analyzed using Analyst 4.1 software (AB Sciex).

2. Standard curves are generated based on linear regression of the analyte/IS peak area ratios (*y*) versus analyte concentration (*x*) using MRMs provided in Table 4.

3. Typically, coefficient of correlation is >0.99.

4. Runs are accepted if calculated controls fall within two standard deviations of target values.

5. The linearity ranges from 50 to 2500 ng/mL. Any sample exceeding 2500 ng/mL is diluted with negative plasma and re-run.

6. Between and with-in run imprecision are <10 %.

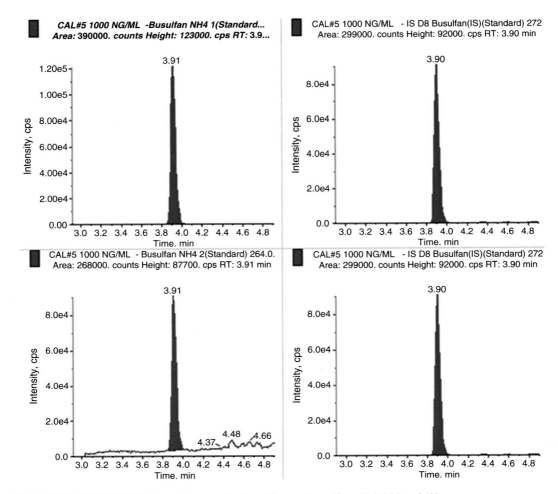

**Fig. 1** HPLC-MS/MS MRM chromatograms for busulfan and busulfan-d8 (1000 ng/mL)

7. Carry-over monitoring is evaluated by injecting negative sample after highest calibrator.

8. Ion suppression is monitored by comparing peak area counts of samples with plasma matrix-free sample and is typically <20 %.

9. Typical chromatograms for busulfan and busulfan-d8 are given in Fig. 1.

### 3.3 Pharmacokinetic Modeling

The data are curve fit using a peeling algorithm to generate initial polyexponential parameter estimates with final parameter estimates determined from an iterative, nonlinear weighted least squares regression algorithm with reciprocal ($1/y2calc$) weighting. Model-dependent pharmacokinetic parameters are calculated from final polyexponential parameter estimates. Alternatively, a model-independent approach can be applied to analyze the data. Area under the plasma concentration versus time curve during the sampling period (AUC0-n) can be calculated using the trapezoidal rule. Extrapolation of the AUC to infinity (AUC0-∞) is calculated by summation of AUC0-n + Cn/λz, where Cn represents the final plasma concentration and λz is the apparent terminal elimination rate constant.

Dose adjustments are driven by clinician defined exposure estimates, typically a desired average steady-state concentration over the entire dosing regimen (Css avg) expressed in ng/mL, or an average AUC over the entire dosing regimen expressed in μmol min. Representative plasma concentration versus time profiles observed with a 6-h and a 24-h dosing interval are illustrated in Fig. 2.

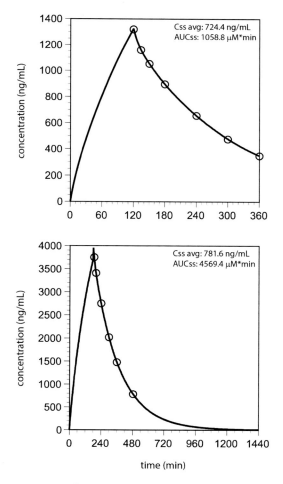

**Fig. 2** Plasma concentration versus time profiles for 6-h and a 24-h dosing interval

## 4   Notes

1. Samples that are clotted, hemolysed, or collected in gel tubes are not suitable.

2. Tertiary standard also serves as calibrator 6.

3. Quaternary standard also serves as calibrator 5.

4. Analyze the negative plasma to assure that it is negative for busulfan and any other unanticipated interference.

5. Controls should be prepared separately and independently from calibrators.

6. Avoid touching the sides of the tube when transferring supernatant.

## References

1. Cavo M, Bandini G, Benni M, Gozzetti A, Ronconi S, Rosti G, Zamagni E, Lemoli RM, Bonini A, Belardinelli A et al (1998) High-dose busulfan and cyclophosphamide are an effective conditioning regimen for allogeneic bone marrow transplantation in chemosensitive multiple myeloma. Bone Marrow Transplant 22:27–32

2. O'Donnell MR, Long GD, Parker PM, Niland J, Nademanee A, Amylon M, Chao N, Negrin RS, Schmidt GM, Slovak ML et al (1995) Busulfan/cyclophosphamide as conditioning regimen for allogeneic bone marrow transplantation for myelodysplasia. J Clin Oncol 13:2973–2979

3. Shah AJ, Lenarsky C, Kapoor N, Crooks GM, Kohn DB, Parkman R, Epport K, Wilson K, Weinberg K (2004) Busulfan and cyclophosphamide as a conditioning regimen for pediatric acute lymphoblastic leukemia patients undergoing bone marrow transplantation. J Pediatr Hematol Oncol 26:91–97

4. Slattery JT, Risler LJ (1998) Therapeutic monitoring of busulfan in hematopoietic stem cell transplantation. Ther Drug Monit 20:543–549

5. Athanasiadou I, Angelis YS, Lyris E, Archontaki H, Georgakopoulos C, Valsami G (2014) Gas chromatographic-mass spectrometric quantitation of busulfan in human plasma for therapeutic drug monitoring: a new on-line derivatization procedure for the conversion of busulfan to 1,4-diiodobutane. J Pharm Biomed Anal 90:207–214

6. Courtney JB, Harney R, Li Y, Lundell G, McMillin GA, Agarwal G, Juenke JM, Mathew A, Gonzalez-Espinoza R, Fleisher M et al (2009) Determination of busulfan in human plasma using an ELISA format. Ther Drug Monit 31:489–494

7. Embree L, Burns RB, Heggie JR, Phillips GL, Reece DE, Spinelli JJ, Hartley DO, Hudon NJ, Goldie JH (1993) Gas-chromatographic analysis of busulfan for therapeutic drug monitoring. Cancer Chemother Pharmacol 32:137–142

8. French D, Sujishi KK, Long-Boyle JR, Ritchie JC (2014) Development and validation of a liquid chromatography-tandem mass spectrometry assay to quantify plasma busulfan. Ther Drug Monit 36:169–174

9. Hassan M, Ehrsson H (1983) Gas chromatographic determination of busulfan in plasma with electron-capture detection. J Chromatogr 277:374–380

10. Juenke JM, Miller KA, McMillin GA, Johnson-Davis KL (2011) An automated method for supporting busulfan therapeutic drug monitoring. Ther Drug Monit 33:315–320

11. Kellogg MD, Law T, Sakamoto M, Rifai N (2005) Tandem mass spectrometry method for the quantification of serum busulfan. Ther Drug Monit 27:625–629

12. Lai WK, Pang CP, Law LK, Wong R, Li CK, Yuen PM (1998) Routine analysis of plasma busulfan by gas chromatography-mass fragmentography. Clin Chem 44:2506–2510

13. Moon SY, Lim MK, Hong S, Jeon Y, Han M, Song SH, Lim KS, Yu KS, Jang IJ, Lee JW et al (2014) Quantification of human plasma-busulfan concentration by liquid chromatography-tandem mass spectrometry. Ann Lab Med 34:7–14

14. Murdter TE, Coller J, Claviez A, Schonberger F, Hofmann U, Dreger P, Schwab M (2001)

Sensitive and rapid quantification of busulfan in small plasma volumes by liquid chromatography-electrospray mass spectrometry. Clin Chem 47:1437–1442

15. Quernin MH, Duval M, Litalien C, Vilmer E, Aigrain EJ (2001) Quantification of busulfan in plasma by liquid chromatography-ion spray mass spectrometry. Application to pharmacokinetic studies in children. J Chromatogr B Biomed Sci Appl 763:61–69

16. Vassal G, Re M, Gouyette A (1988) Gas chromatographic-mass spectrometric assay for busulfan in biological fluids using a deuterated internal standard. J Chromatogr 428:357–361

# Chapter 10

# High-Throughput Quantitation of Busulfan in Plasma Using Ultrafast Solid-Phase Extraction Tandem Mass Spectrometry (SPE-MS/MS)

## Loralie J. Langman, Darlington Danso, Enger Robert, and Paul J. Jannetto

## Abstract

Busulfan is a commonly used antineoplastic agent to condition/ablate bone marrow cells before hematopoietic stem cell transplant. While intravenous (IV) formulations of busulfan are now available and have lower incidences of toxicity and treatment related mortality compared to oral dosing, it still displays large pharmacokinetic variability. As a result, studies have shown that therapeutic drug monitoring is clinically useful to minimize graft failure, disease reoccurrence, and toxicities like veno-occlusive disease and neurologic toxicity. Current methods for assaying busulfan include the use of GC/MS, HPLC, and LC-MS/MS. The clinical need for faster turnaround times and increased testing volumes has required laboratories to develop faster methods of analysis for higher throughput of samples. Therefore, we present a method for the quantification of busulfan in plasma using an ultrafast SPE-MS/MS which has much faster sample cycle times (<20 s per sample) and comparable analytical results to GC/MS.

**Key words** Busulfan, SPE-MS/MS AUC, Pharmacokinetic monitoring

## 1 Introduction

Busulfan (1, 4-butanediol dimethansulfonate) was approved by the US Food and Drug Administration (FDA) in 1999 for the treatment of chronic myelogenous leukemia; it is also used in combination with other drugs as a conditioning agent prior to bone marrow transplantation [1]. Busulfan is a bifunctional antineoplastic whose mechanism of action includes alkylation and cross-linking of strands of DNA to prevent its replication [2]. Specifically, busulfan prefers to react with the $N_7$ nitrogen of guanine or the $N_3$ nitrogen of adenine [3, 4]. The functionality of busulfan allows it to form an intra-strand cross-link between adjacent nucleotides in a DNA chain [5, 6] which is accomplished by the $SN_2$ reactions with the nucleotides on both of the functionalized carbons on the busulfan.

Uttam Garg (ed.), *Clinical Applications of Mass Spectrometry in Drug Analysis: Methods and Protocols*, Methods in Molecular Biology, vol. 1383, DOI 10.1007/978-1-4939-3252-8_10, © Springer Science+Business Media New York 2016

The intra-strand crosslinks formed are believed to be responsible for busulfan's cytotoxicity and are thought to inhibit DNA replication leading to eventual cell death (apoptosis). To minimize cytotoxicity while ensuring adequate concentrations of busulfan completely destroy the bone marrow, therapeutic drug monitoring is critical. Studies have shown that monitoring the busulfan area under the plasma drug concentration-time curve (AUC) and steady-state concentrations (Css) have been related to therapeutic outcome [7].

Busulfan is commonly administered intravenously through a central venous catheter. It is given as a 2-h infusion, every 6 h for four consecutive days resulting in a total of 16 doses [8]. The usual adult dose is 0.8 mg/kg of ideal body weight or actual body weight. Intravenous dosing is usually guided by pharmacokinetic evaluation of the area under the curve (AUC) and clearance after the first dose [9]. Busulfan AUC and clearance are calculated after the quantification of busulfan concentration in plasma collected immediately after the termination of a 2-h intravenous infusion of 0.8 mg/kg busulfan, as well as 1 h, 2 h, and 4 h after the end of infusion. The AUC is calculated using the trapezoidal rule. The optimal dose is calculated based on the assumption that the ideal AUC is 1100 µM*min. AUC greater than 1500 µM*min is associated with hepatic veno-occlusive disease [10], and AUC less than 900 µM*min indicates incomplete bone marrow ablation. In order to facilitate monitoring, there is the need for fast turn–around-times (TAT) for the analysis of busulfan in plasma. Therefore, an ultrafast SPE/MS/MS method with a cycle time of 20 s per injection for the quantitation of busulfan in plasma is presented.

## 2 Materials

### 2.1 Samples

Heparinized plasma samples must be collected, placed on ice, and centrifuged as quickly as possible to prevent degradation [11] (see **Note 1**). However, once separated, samples stored refrigerated were stable up to 3 days and samples stored frozen (–70 °C) with up to five freeze-thaw cycles were stable up to 28 days. Following 24 h of storage at ambient temperature, the drug concentrations decrease significantly and was therefore considered not stable under ambient conditions [12].

### 2.2 Reagents

1. Wash solution #1(clinical laboratory reagent water (CLRW) with 0.1 % formic acid): In a 2 L reagent bottle, add 2 L CLRW. Add 2 mL formic acid and mix well. It is stable for 1 month at 20–27 °C.

2. Wash solution #2 (90 % acetonitrile with 10 % isopropanol): In a 2 L reagent bottle, add 1.8 L acetonitrile. Add 200 mL isopropanol and mix well. It is stable for 1 year at 20–27 °C.

3. Mobile phase 1/reconstitution solution: In a 2.0 L reagent bottle add 2.0 L CLRW. Add 1.54 g ammonium acetate and mix well. Add 2.0 mL formic acid and mix well. Add 180 µL TFA and mix well. It is stable 1 month at 20–27 °C.

4. Mobile phase 2 (50 % methanol/CLRW): In a 2.0 L reagent bottle add 1.0 L methanol. Add 1.0 L CLRW and mix well. It is stable 2 months at 20–27 °C.

5. Mobile phase 3: In a 2 L reagent bottle, add 2.0 L methanol. Add 1.54 g ammonium acetate and mix well. Add 2 mL formic acid and mix well. Add 180 µL TFA and mix well. It is stable 1 year at 20–27 °C.

**2.3 Standards and Calibrators**

1. Busulfan, Stock I (Sigma-Aldrich).

2. Busulfan Stock II Standard (1.0 mg/mL): Accurately weigh out 10 mg of busulfan and transfer to a 10 mL volumetric flask. Bring to volume with acetone and mix thoroughly. Store at −10 to −35 °C protected from light. Stable for 1 year after preparation or until expiration of Stock 1 Standard, whichever comes first.

3. Busulfan Intermediate Standard (100 µg/mL): Transfer 2.5 mL of Stock II standard solution (1.0 mg/mL) to a 25 mL volumetric flask. Bring to volume with methanol and mix thoroughly. Store at −10 to −35 °C protected from light. Stable for 1 year after preparation or until expiration of Stock II Standards, whichever comes first.

4. Calibrators are prepared according to Table 1.

**Table 1**
**Preparation of calibrators: the total volume is made to 50 mL with drug-free plasma**

| Concentration (ng/mL) | µL of working standard (100 µg/mL) |
| --- | --- |
| 25 | 12.5 |
| 200 | 100 |
| 1000 | 500 |
| 3000 | 1500 |
| | *µL of working standard (1 mg/mL)* |
| 7500 | 375 |

Calibrators are stable for 1 year at −20 °C

**Table 2**
**Preparation of quality control: the total volume is made to 50 mL with drug-free plasma**

| Concentration (ng/mL) | μL of working standard (100 μg/mL) |
|---|---|
| 300 | 150 |
| 2500 | 1250 |
| | *μL of working standard (1000 μg/mL)* |
| 5000 | 250 |

Quality controls are stable for 1 year at –20 °C

*2.4 Quality Controls and Internal Standard*

1. Quality controls samples are prepared in house according to the Table 2.

2. Primary internal standard: Busulfan-d$_4$—Prepared using the method of Vassal and Gouyette [13] or can be bought commercially (Cambridge Isotope Laboratories).

3. Busulfan-d$_4$ Stock II Internal Standard, 1.0 mg/mL: Accurately weigh out 10 mg of busulfan internal standard and transfer to a 10 mL volumetric flask. Bring to volume with ethyl acetate and mix thoroughly. Store at –10 to –35 °C protected from light. Stable for 1 year after preparation or until expiration of Stock I Standard, whichever comes first.

4. Busulfan-d$_4$ Intermediate Internal Standard (100.0 μg/mL): Transfer 1 mL of Stock II standard (1.0 mg/mL) to a 10 mL volumetric flask. Bring to volume with methanol and mix thoroughly. Store at –10 to –35 °C protected from light. Stable for 1 year after preparation or until expiration of Stock II Standards, whichever comes first.

5. Busulfan-d$_4$ Working Internal Standard (2.0 μg/mL): Transfer 2 mL of the 100 μg/mL Intermediate Internal Standard to a 100 mL volumetric flask. Bring to volume with methanol and mix thoroughly. Transfer solution into ten amber vials and seal with screw caps utilizing a rubber/Teflon septum. Store at –10 to –35 °C. Stable for 1 year after preparation or until expiration of Stock II Standards, whichever comes first.

*2.5 Supplies and Equipment*

1. 96 Deep-well 1.5 mL plates (Nalgen Nunc).

2. SPE cartridge C18 (Agilent RapidFire).

3. Borosilicate glass test tubes; 16 × 125 mm and 16 × 100 mm.

4. Agilent 1200 series HPLC using Agilent RapidFire liquid chromatography.

5. Agilent 6400 series QQQ mass spec detector.

# 3   Methods

### 3.1  Stepwise Procedure

1. Aliquot 100 μL of each standard, control, and patient sample into individually labeled $16 \times 125$ mm glass tubes.

2. To each standard, control, and patient sample add:

    (a)  50 μL of working internal standard.

    (b)  3.0 mL of *n*-butyl chloride.

3. Vortex for 2 min on multi-vortexer.

4. Centrifuge at $2100 \times g$ for 5 min.

5. Transfer organic supernatant to individually labeled $16 \times 100$ glass tubes.

6. Dry down under a gentle stream of nitrogen at $\leq 40$ °C.

7. Add 600 μL of reconstitution solution.

8. Vortex for 10 s.

9. Transfer extracts to the 96 deep-well plate.

10. Inject 10 μL of each extract on a RapidFire-MS/MS system.

### 3.2  Data Analysis

1. Samples were analyzed at the rate of 20 s per sample. Analytical conditions are described in Table 3.

2. MassHunter Triple Quadruple Acquisition software (B.04.01) with Qualitative Analysis (B.04.00) and Quantitative Analysis (B.04.00), and RapidFire Integrator Software.

**Table 3**
**Analytical conditions for HPLC-MS-MS**

| RapidFire cycle conditions | | | |
|---|---|---|---|
| State 1 | Aspirate | 600 ms | |
| State 2 | Load/wash | 3500 ms | Mobile phase 1/ mobile phase 2 |
| State 3 | Elute | 3500 ms | Mobile phase 3 |
| State 4 | Re-equilibrate | 1000 ms | Mobile phase 1 |
| Triple quadrupole conditions | | | |
| Gas temp | 220 °C | | |
| Gas flow | 19 L/min | | |
| Nebulizer | 50 psi | | |
| Sheath gas temp | 250 °C | | |
| Sheath gas flow | 11 L/min | | |
| Capillary | 3000 V | | |
| Nozzle voltage | 0 V | | |

**Table 4**
**MRMs for busulfan and busulfan-D4**

| Analyte | Precursor ion | Product ion | CE (v) |
|---|---|---|---|
| Busulfan | 264.1 | 151.1 | 1 |
| Busulfan qualifier | 264.1 | 55.1 | 10 |
| Busulfan-D4 | 268.1 | 155.1 | 0 |
| Busulfan-D4 qualifier | 268.1 | 59.1 | 12 |

3. The ions used for identification and quantification are listed in Table 4.

4. The linearity/limit of quantitation of the method is 25–7500 ng/mL. Samples in which the drug concentrations exceed the upper limit of quantitation should be diluted with drug free plasma and retested.

5. A typical calibration curve has correlation coefficient ($r^2$) of >0.99.

6. Typical intra and inter-assay imprecision throughout the analytical range is <5 %.

7. Quality control: The analytical run is considered acceptable if the calculated concentrations of drugs in the controls are within ±10 % of target values. The quantifying ion in the sample is considered acceptable if the ratios of qualifier ions to quantifying ion are within ±20 % of the ion ratios for the calibrators.

8. A typical Chromatogram is shown in Fig. 1.

### 3.3 Reporting

1. The units, the AUC should be reported in μM*min, the clearance in (mL/min)/Kg, and dose in mg.

2. Dosing is usually guided by pharmacokinetic evaluation of the area under the curve (AUC) and clearance calculation of AUC is performed using the trapezoidal rule, and clearance is the dose divided by the AUC.

## 4    Notes

1. This step is particularly important to prevent degradation of the drug in whole blood and the possibility of reporting out a falsely decreased busulfan concentration.

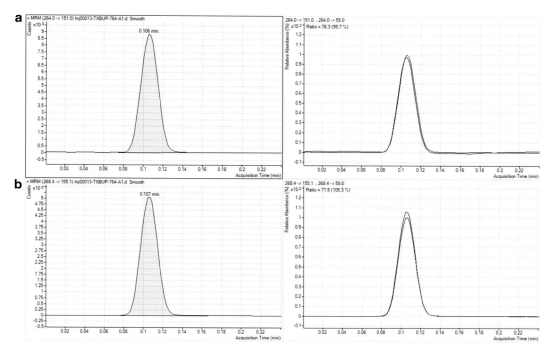

**Fig. 1** Chromatogram of (**a**) busulfan at the LOQ (25 ng/mL – Signal to noise ratio 1302), and (**b**) Busulfan-d4 Internal Standard

## References

1. Buggia I, Locatelli F, Regazzi MB, Zecca M (1994) Busulfan. Ann Pharmacother 28:1055–1062

2. Kohn KW (1996) Beyond DNA cross-linking: history and prospects of DNA-targeted cancer treatment--fifteenth Bruce F. Cain Memorial Award Lecture. Cancer Res 56:5533–5546

3. Brookes P, Lawley PD (1960) The reaction of mustard gas with nucleic acids in vitro and in vivo. Biochem J 77:478–484

4. Kohn KW, Hartley JA, Mattes WB (1987) Mechanisms of DNA sequence selective alkylation of guanine-N7 positions by nitrogen mustards. Nucleic Acids Res 15:10531–10549

5. Hurley LH (2002) DNA and its associated processes as targets for cancer therapy. Nat Rev Cancer 2:188–200

6. Newbold RF, Warren W, Medcalf AS, Amos J (1980) Mutagenicity of carcinogenic methylating agents is associated with a specific DNA modification. Nature 283:596–599

7. Slattery JT, Risler LJ (1998) Therapeutic monitoring of busulfan in hematopoietic stem cell transplantation. Ther Drug Monit 20:543–549

8. GlaxoSmithKline, Research Triangle Park, NC 27709. 2003

9. Baselt RC. Busulfan. In: Baselt RC, Ed. Disposition of toxic drugs and chemical in man. 9th ed. Foster, City, CA: Biomedical Publications; 2011. p. 218–220.

10. Slattery JT, Sanders JE, Buckner CD, Schaffer RL, Lambert KW, Langer FP, Anasetti C, Bensinger WI, Fisher LD, Appelbaum FR et al (1995) Graft-rejection and toxicity following bone marrow transplantation in relation to busulfan pharmacokinetics. Bone Marrow Transplant 16:31–42

11. Moon SY, Lim MK, Hong S, Jeon Y, Han M, Song SH, Lim KS, Yu KS, Jang IJ, Lee JW et al (2014) Quantification of human plasma-busulfan concentration by liquid chromatography-tandem mass spectrometry. Ann Lab Med 34:7–14

12. Danso D, Jannetto PJ, Enger R, Langman LJ (2015) High-throughput validated method for the quantitation of busulfan in plasma using ultrafast SPE-MS/MS. Ther Drug Monit 37:319–324

13. Vassal G, Re M, Gouyette A (1988) Gas chromatographic-mass spectrometric assay for busulfan in biological fluids using a deuterated internal standard. J Chromatogr 428:357–361

# Chapter 11

# Quantification of 11-Carboxy-Delta-9-Tetrahydrocannabinol (THC-COOH) in Meconium Using Gas Chromatography/Mass Spectrometry (GC/MS)

## Judy Peat, Brehon Davis, Clint Frazee, and Uttam Garg

## Abstract

Maternal substance abuse is an ongoing concern and detecting drug use during pregnancy is an important component of neonatal care when drug abuse is suspected. Meconium is the preferred specimen for drug testing because it is easier to collect than neonatal urine and it provides a much broader time frame of drug exposure. We describe a method for quantifying 11-carboxy-delta-9-tetrahydrocannabinol (THC-COOH) in meconium. After adding a labeled internal standard (THC-COOH D9) and acetonitrile, samples are sonicated to release both free and conjugated THC-COOH. The acetonitrile/aqueous layer is removed and mixed with a strong base to hydrolyze the conjugated THC-COOH. The samples are then extracted with an organic solvent mixture as part of a sample "cleanup." The organic solvent layer is discarded and the remaining aqueous sample is acidified. Following extraction with a second organic mixture, the organic layer is removed and concentrated to dryness. The resulting residue is converted to a trimethylsilyl (TMS) derivative and analyzed using gas chromatography/mass spectrometry (GC/MS) in selective ion monitoring (SIM) mode.

**Key words** Substance abuse, Meconium, Marijuana, 11-Carboxy-delta-9-tetrahydrocannabinol, Carboxy-THC

## 1 Introduction

Illicit drugs use during pregnancy remains a significant concern, and is associated with adverse fetal and maternal outcome. Amongst abused substances, cannabis remains the most commonly abused in the United States [1]. Various methods such as interviewing the mother in person or by questionnaire and drug testing in different specimen matrices are used to determine prenatal drug exposure [2–5]. Due to the legal repercussions of admitting illicit drug use, self-reported drug use is not reliable [2–4]. Urine from mother or infant is typically positive only for few days after the drug exposure. Meconium is a preferred sample to determine fetal drug exposure as it can provide maternal drug abuse history for several months

Uttam Garg (ed.), *Clinical Applications of Mass Spectrometry in Drug Analysis: Methods and Protocols*, Methods in Molecular Biology, vol. 1383, DOI 10.1007/978-1-4939-3252-8_11, © Springer Science+Business Media New York 2016

because it begins forming between the 12th and 16th weeks of gestation, and it accumulates until shortly after birth. Meconium is a gelatinous, heterogeneous substance comprised of epithelial and squamous cells and amniotic fluid, swallowed by the fetus during the last half of pregnancy, and voided as first stools following birth. It is hypothesized that the fetus excretes drug into bile and amniotic fluid, and then the drug accumulates in meconium by direct disposition or by swallowing amniotic fluid [2, 3].

Because meconium is a thick and heterogeneous material, it is a difficult sample to work with, and requires special preparation before drug extraction. In general meconium is homogenized in an organic solvent for drug extraction. The extract is either used directly or dried and reconstituted in an aqueous buffer, and tested by immunoassay or mass spectrometric methods. Immunoassay positive results should be confirmed by a mass spectrometry method. Both gas and liquid chromatography mass spectrometric methods have been described in the literature [6–9]. We describe a GC/MS method for measuring total THC-COOH levels in meconium. The method is simple and reproducible, and has a linear range of 10–500 ng/g.

## 2 Materials

### 2.1 Sample

1 g meconium.

### 2.2 Solvents and Reagents

1. Bis-(trimethylsilyl)trifluoroacetamide (BSTFA) with 1 % trimethylchlorosilane (TMCS) (United Chemical Technologies, Bristol, PA).

2. 11.8 N Potassium hydroxide: Add approximately 500 mL of deionized water to a 1 L volumetric flask. Slowly add 662 g of KOH pellets and bring the volume to 1 L with deionized water. Store in an amber bottle. Stable for 1 year at room temperature.

3. Hexanes: Ethyl acetate (8:2): Combine 800 mL hexanes with 200 mL of ethyl acetate. Store in an amber bottle. Stable for 1 year at room temperature.

4. 0.1 M acetic acid: Add approximately 400 mL of deionized water to a 500 mL volumetric flask. Slowly add 2.87 mL glacial acetic acid and bring the volume to 500 mL with deionized water. Stable for 6 months at room temperature.

5. 0.2 N Sodium hydroxide: Add 10 mL 1.0 N NaOH to a 50 mL volumetric flask and bring the volume to 50 mL with deionized water. Stable for 6 months at room temperature.

### 2.3 Standards

1. Primary standard: 100 μg/mL THC-COOH (Cerilliant).

2. Primary internal standard: 100 μg/mL THC-COOH D9 (Cerilliant).

**Table 1**
**Preparation of calibrators and controls**

| Calibrator/control concentration (ng/g) | µL of working tertiary standard | µL of working secondary standard |
|---|---|---|
| 15 | 15 | |
| 20 (control) | 20 | |
| 50 | 50 | |
| 100 | 100 | |
| 500 | | 50 |

The 20 ng/g control is made from a separate standard

3. Working secondary standard, 10 µg/mL THC-COOH: Add 1 mL primary standard to a 10 mL volumetric flask and bring the volume to 10 mL with methanol. Stable for 1 year at –20 °C.

4. Working tertiary standard, 1 µg/mL THC-COOH: Add 1 mL of working secondary standard to a 10 mL volumetric flask and bring the volume to 10 mL with methanol. Stable for 1 year at –20 °C.

5. Working internal standard, 2 µg/mL THC-COOH D9: Add 1 mL primary internal standard to 50 mL volumetric flask and bring the volume to 50 mL with methanol. Stable for 1 year at –20 °C.

*2.4 Calibrators and Controls*

1. Prepare working calibrators and controls according to Table 1 by adding the indicated tertiary or secondary standard volume to extraction tubes which have been pre-coated with 1 g negative meconium (*see* **Note 1**).

2. In-house meconium controls: Bio-Rad (Bio-Rad Laboratories) THC-COOH urine controls (two levels) were used to prepare THC-COOH meconium controls. 1 mL urine control was added to 1 g negative meconium and vortexed to mix.

*2.5 Analytical Supplies*

1. 16 × 100 screw-cap glass tubes for extraction.

2. 13 × 100 screw-cap glass tubes for extract concentration.

3. Transfer pipets (Samco Scientific, San Fernando CA).

4. Auto sampler vials (12 × 32 mm with crimp caps) with 0.3 mL limited volume inserts (P.J. Cobert Associates, St. Louis, MO).

5. GC column: Zebron ZB-1 with dimensions of 15 m × 0.25 mm × 0.25 µm (Phenomenex, Torrance, California).

6. Plain wood applicators: These are used to evenly spread the meconium around the glass tube (Fisher Scientific, Waltham, MA, USA).

**2.6  Equipment**

1. A gas chromatograph/mass spectrometer system (GC/MS; 6890/5975 or 5890/5972) with autosampler and operated in electron impact mode (Agilent Technologies, Wilmington, DE).

2. TurboVap®IV Evaporator (Zymark Corporation, Hopkinton, MA, USA).

# 3  Method

**3.1  Stepwise Procedure**

1. Weigh out 1 g of each patient meconium into a $16 \times 100$ mm test tube. Record weight to within two decimal places. Spread meconium as evenly as possible onto the sides of the tube for a uniform thin coating of sample. Freeze until analysis, at least overnight (*see* **Note 2**).

2. For each of the four calibrators, the blank (negative control) and the three controls, weigh out 1 g of negative meconium into appropriately labeled $16 \times 100$ mm test tubes. Spread meconium evenly onto the sides of the tube. Add working THC-COOH for each calibrator and the 20 ng/g in-house control (*see* Table 1). Add 1 mL of each control, prepared from Bio-Rad controls, to appropriately labeled tubes. Cap and vortex to mix. Freeze all meconium specimens until analysis (at least overnight) and thaw for 15 min at room temperature before analysis.

3. Prepare an unextracted standard by adding 100 µL working THC-COOH tertiary standard and 100 µL working THC-COOH D9 internal standard to a concentration tube. Set aside until **step 18**.

4. Add 4 mL acetonitrile to each tube.

5. Add 100 µL of working THC-COOH D9 IS to each tube. Cap and vortex to mix. Sonicate tubes (using a beaker or test tube rack) for 5 min. Centrifuge for 5 min at $1200 \times g$.

6. Transfer organic layer to appropriately labeled clean concentration tubes.

7. Spread meconium around the sides of the original extraction tube as much as possible for a uniform thin coating of sample.

8. Add 2 mL of acetonitrile to the original sample tubes for a second extraction. Cap and vortex to mix. Sonicate for 5 min. Centrifuge for 5 min at $1200 \times g$.

9. Add the 2 mL organic to the concentration tube containing the first 4 mL acetonitrile extract.

10. Concentrate the combined organic extract to less than 1 mL under nitrogen at 40 °C (*see* **Note 3**).

11. Add 2 mL 0.2 N NaOH to each tube.

12. Add 100 µL 11.8 N KOH to each tube. Vortex. Let sit for a minimum of 15 min.

13. Add 5 mL hexane:ethyl acetate (8:2) to each tube. Cap and rock for a minimum of 15 min. Centrifuge for 5 min at $1200 \times g$.

14. Discard upper organic layer. To the bottom aqueous layer add 2 mL 0.1 M acetic acid.

15. Add 200 µL glacial acetic acid.

16. Add 3 mL hexane:ethyl acetate (8:2). Cap and rock for 15 min. Centrifuge for 5 min at $1200 \times g$.

17. Transfer upper organic layer to a clean concentration tube.

18. Concentrate to dryness under nitrogen at 40 °C.

19. Reconstitute with 100 µL BSTFA + TMCS.

20. Cap and incubate for 10 min at 65 °C in heating block.

21. Cool and transfer to appropriately labeled autosampler vials.

22. Inject 1 µL onto GC/MS for analysis (GC-MS operating condition are given in Table 2).

**3.2  Data Analysis**

1. Data are analyzed using Target Software (Thru-Put Systems, Orlando, FL) or similar software.

2. Standard curves are generated based on linear regression of the analyte/IS peak area ratio ($y$) versus analyte concentration ($x$) using the quantifying ion listed in Table 3.

**Table 2**
**GC-MS operating conditions**

| | |
|---|---|
| Oven program | 120 °C for 0.5 min<br>Then 30 °C/min to 280 °C<br>Hold for 6 min |
| Front inlet | Mode: splitless<br>Injection temperature: 250 °C<br>Column pressure: 5 psi<br>Purge time on at 1 min |
| Mass spectrometer | Mode: Electron impact at 70 eV<br>Detector temperature: 280 °C |

**Table 3**
**Quantification and qualifier ions for THC-COOH and THC-COOH D9**

| | Quantitation ions | Qualifier ions |
|---|---|---|
| THC-COOH | 473 | 371,488 |
| THC-COOH D9 | 479 | 380,497 |

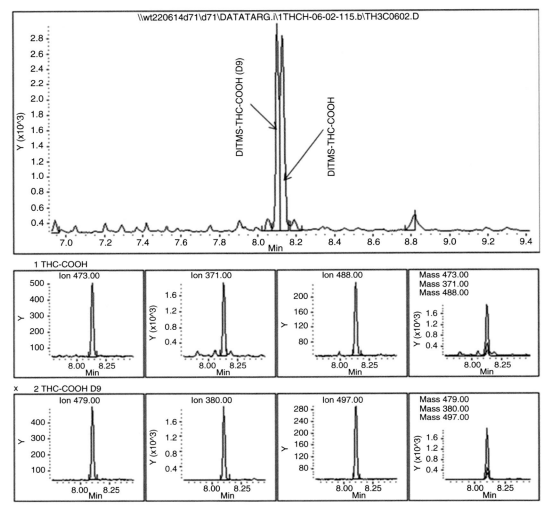

**Fig. 1** GC-MS chromatogram of TMS derivatives of THC-COOH and THC-COOH-D9 (100 ng/g). The bottom panels show selected ion chromatograms of THC-COOH and THC-COOH-D9 TMS derivatives

3. Typical total and SIM chromatogram are shown in Fig. 1.

4. Analytical run is considered acceptable if the control values are within 20 %.

5. Typical coefficient of correlation is >0.99.

6. Linearity of the method is from 10 to 500 ng/g.

7. Typical intra- and inter-assay imprecision is <10 %.

## 4  Notes

1. Calibrators and controls are prepared independently.

2. Freezing the meconium specimen overnight at −20 °C increases extraction recovery.

3. Concentration of organic extract takes ~30 min.

### References

1. http://www.monitoringthefuture.org/, Accessed 6/11/15

2. Ostrea EM Jr, Brady MJ, Parks PM, Asensio DC, Naluz A (1989) Drug screening of meconium in infants of drug-dependent mothers: an alternative to urine testing. J Pediatr 115:474–477

3. Ostrea EM Jr, Knapp DK, Tannenbaum L, Ostrea AR, Romero A, Salari V, Ager J (2001) Estimates of illicit drug use during pregnancy by maternal interview, hair analysis, and meconium analysis. J Pediatr 138:344–348

4. Wingert WE, Feldman MS, Kim MH, Noble L, Hand I, Yoon JJ (1994) A comparison of meconium, maternal urine and neonatal urine for detection of maternal drug use during pregnancy. J Forensic Sci 39:150–158

5. Kwong TC, Ryan RM (1997) Detection of intrauterine illicit drug exposure by newborn drug testing. National Academy of Clinical Biochemistry. Clin Chem 43:235–242

6. Coles R, Clements TT, Nelson GJ, McMillin GA, Urry FM (2005) Simultaneous analysis of the Delta9-THC metabolites 11-nor-9-carboxy-Delta9-THC and 11-hydroxy-Delta9-THC in meconium by GC-MS. J Anal Toxicol 29:522–527

7. Tynon M, Porto M, Logan BK (2015) Simplified analysis of 11-hydroxy-delta-9-tetrahydrocannabinol and 11-carboxy-delta-9-tetrahydrocannabinol in human meconium: method development and validation. J Anal Toxicol 39:35–40

8. Gray TR, Shakleya DM, Huestis MA (2009) A liquid chromatography tandem mass spectrometry method for the simultaneous quantification of 20 drugs of abuse and metabolites in human meconium. Anal Bioanal Chem 393:1977–1990

9. Ristimaa J, Gergov M, Pelander A, Halmesmaki E, Ojanpera I (2010) Broad-spectrum drug screening of meconium by liquid chromatography with tandem mass spectrometry and time-of-flight mass spectrometry. Anal Bioanal Chem 398:925–935

# Chapter 12

# Quantitation of Carisoprodol and Meprobamate in Urine and Plasma Using Liquid Chromatography-Tandem Mass Spectrometry (LC-MS/MS)

## Matthew H. Slawson and Kamisha L. Johnson-Davis

## Abstract

Carisoprodol and meprobamate are centrally acting muscle relaxant/anxiolytic drugs that can exist in a parent–metabolite relationship (carisoprodol → meprobamate) or as a separate pharmaceutical preparation (meprobamate aka Equanil, others). The monitoring of the use of these drugs has both clinical and forensic applications in pain management applications and in overdose situations. LC-MS/MS is used to analyze urine or plasma/serum extracts with deuterated analogs of each analyte as internal standards to ensure accurate quantitation and control for any potential matrix effects. Positive ion electrospray is used to introduce the analytes into the mass spectrometer. Selected reaction monitoring of two product ions for each analyte allows for the calculation of ion ratios which ensures correct identification of each analyte, while a matrix-matched calibration curve is used for quantitation.

**Key words** Carisoprodol, Meprobamate, Muscle relaxant, Mass spectrometry

## 1 Introduction

Carisoprodol is a centrally acting muscle relaxant marketed under the brand name Soma. It has a pharmacologically active, N-dealkylated metabolite, meprobamate (aka Miltown, Equanil, Meprospan) that can be separately prescribed as an anxiolytic although this therapeutic use has been supplanted by the benzodiazepines. Both drugs are Schedule IV controlled substances with abuse potential. These compounds are therapeutically monitored for compliance [1], and to maintain patients in the therapeutic range [2]. They are also analyzed in acute intoxication cases, since they can produce CNS depression, and particularly in combination with other CNS depressants, may be life-threatening [3–5]. Carisoprodol is metabolized to meprobamate, which can then be hydroxylated and glucuronidated. Less than 1 % of a dose is excreted as carisoprodol in a 24 h urine; 5 % as meprobamate. Meprobamate is metabolized to hydroxymeprobamate, an inactive

Uttam Garg (ed.), *Clinical Applications of Mass Spectrometry in Drug Analysis: Methods and Protocols*, Methods in Molecular Biology, vol. 1383, DOI 10.1007/978-1-4939-3252-8_12, © Springer Science+Business Media New York 2016

metabolite, and is also glucuronidated. Approximately 5 % is eliminated in the urine unchanged, a comparable amount of hydroxy metabolite is excreted, and as much as 65 % of the dose is eliminated in 48 h in the urine as the glucuronide conjugate [6]. Several methods have been published for the analysis of carisoprodol and meprobamate in a variety of matrices [7–11].

This assay utilizes HPLC-MS/MS to measure both carisoprodol and meprobamate in urine or plasma/serum and can be used as the confirmation assay for an immunoassay screen.

Supported liquid extraction (SLE+, Biotage) is used for sample clean-up. The final preparation is mixed well and injected onto the LC-MS/MS. Qualitative identification is made using unique MS/MS transitions, ion ratios of those transitions, and chromatographic retention time. Quantitation is performed using a daily calibration curve of prepared calibration samples and using peak area ratios of analyte to internal standard to establish the calibration model. Patient sample concentrations are calculated based on the calibration model's mathematical equation. Quantitative accuracy is monitored with QC samples independently prepared with known concentrations of carisoprodol and meprobamate and comparing the calculated concentration with the expected concentration.

## 2   Materials

### 2.1   Samples

1. Pre-dose (trough) draw at a steady-state concentration for serum/plasma. Separate serum or plasma from cells within 2 h of collection.

2. Random collection for urine.

3. Specimens can be stored for at least 20 days (refrigerated or frozen) prior to analysis.

### 2.2   Reagents

1. Verified negative urine pool.

2. Verified negative serum/plasma pool.

3. 0.5 M ammonium hydroxide: 140 mL concentrated ammonium hydroxide, QS to 500 mL with deionized water. Verify pH at each use and replace solution when needed.

4. 0.1 % Formic Acid in deionized water and methanol (90:10): To 90 mL of 0.1 % formic acid in deionized water, add 10 mL of methanol for a final volume of 100 mL.

5. Mobile Phase A: 0.1 % formic acid in deionized water.

6. Mobile Phase B: 0.1 % formic acid in acetonitrile.

### 2.3   Standards and Calibrators

1. Carisoprodol, 1.0 mg/mL stock standard prepared in methanol (Cerilliant, Round Rock, TX).

2. Meprobamate, 1.0 mg/mL stock standard prepared in methanol (Cerilliant, Round Rock, TX).

**Table 1**
**Preparation of calibrators. For each concentration, the total volume is made to 10 mL with drug-free plasma or human urine**

| Final concentration ng/mL | Solution µL | Stock solution concentration |
|---|---|---|
| 10,000 | 100 | 1 mg/mL |
| 7500 | 750 | 100 ng/µL |
| 5000 | 500 | 100 ng/µL |
| 1000 | 100 | 100 ng/µL |
| 500 | 50 | 100 ng/µL |
| 100 | 10 | 100 ng/µL |

3. Intermediate working solution (carisoprodol and meprobamate at 100 ng/µL in methanol): Add ~3 mL of methanol to a 5 mL volumetric flask. Add 0.5 mL of each calibrator reference material to the flask, QS to 5 mL with methanol, add a stir bar and stopper and mix for 30 min at room temperature. Aliquot as appropriate for subsequent use. Store frozen, stable for 1 year. This volume can be scaled up or down as appropriate.

4. Prepare working calibrators using the solutions described above to prepare 10 mL of each using volumetric glassware. Add approximately 5 mL certified negative urine to a labeled volumetric flask. Add the appropriate volume/concentration as shown in Table 1 of standard material to the flask. QS to 10 mL using certified negative urine. Add a stir bar and stopper and mix for 30 min at room temperature. Aliquot as appropriate for future use. Store aliquots frozen, stable for 1 year. This volume can be scaled up or down as appropriate (*see* **Note 1**).

**2.4 Controls and Internal Standard**

1. Controls: May be purchased from a third party and prepared according to the manufacturer. They can also be prepared in-house independently from calibrators' source material using Table 1 as a guideline (*see* **Note 1**).

2. Internal Standard: Carisoprodol-D$_7$ 1.0 mg/mL in methanol (Cerilliant, Round Rock, TX). Meprobamate-D$_7$ 1.0 mg/mL in methanol (Cerilliant, Round Rock, TX). Add ~80 mL of methanol to a 100 mL volumetric flask. Add 1 mL of each internal standard to the flask, QS to 100 mL with methanol. Add a stir bar and a stopper. Mix for 30 min at room temperature. Aliquot as needed for use in this assay (volumes can be scaled up or down as appropriate). Store frozen, stable for 1 year (*see* **Note 1**).

**2.5 Supplies**

1. Transfer/aliquoting pipettes and tips.

2. Sample preparation tubes, such as microcentrifuge tubes or small culture tubes. These will be used for combining and mixing the specimen aliquot, internal standard and pH buffer prior to adding to the SLE+ cartridge.

3. 400 μL SLE+ array cartridges (Biotage, Charlotte, NC) and array frame (*see* **Note 2**).

4. Instrument compatible autosampler vials with injector appropriate caps or 96 deepwell plate and capping mat.

5. Kinetex 1.7 μm XB-C18 100 Å column size 50×2.1 mm (Phenomenex, Torrance, CA).

6. Security Guard Kit with Security Guard Cartridges C18, 4×2.0 mm ID (Phenomenex, Torrance, CA).

**2.6 Equipment**

1. Vortex mixer.

2. Positive pressure extraction manifold (*see* **Note 3**).

3. Solvent evaporation manifold.

4. Low dead volume binary HPLC pump, thermostatted column compartment with switching valve, vacuum degasser (Agilent, Santa Clara, CA), and DLW CTC PAL auto-sampler (Leap Technologies, Carrboro, NC).

5. API 4000 LC-MS/MS system with Turbo Ion source running current version of Analyst software (ABSciex, Framingham, MA).

# 3 Methods

**3.1 Stepwise Procedure**

1. Briefly vortex or invert each sample to mix.

2. Aliquot 50 μL of each patient sample, calibrator and QC into appropriately labeled micro-centrifuge tubes (or equivalent).

3. Add 5 μL of working internal standard solution to each vial.

4. Add 200 μL 0.5 M ammonium hydroxide to each vial.

5. Cap (if appropriate) each vial and vortex briefly.

6. Assemble the SLE+ array using an appropriate number of cartridge wells into the array plate frame (*see* **Note 3**).

7. Install the assembled array plate onto the positive pressure manifold with a clean collection plate under the SLE+ array plate.

8. Transfer the contents of each specimen tube (from **steps 2–5**) to an SLE+ array well cartridge. Allow the liquid to soak into the SLE+ bed for a minimum of 5 min (*see* **Note 4**).

9. After 5 min of equilibration, add 900 μL of methylene chloride to each well. Allow the contents to drip through the well under gravity (*see* **Note 4**).

10. Add a second 900 µL volume of methylene chloride and allow to pass through the SLE bed under gravity (*see* **Note 5**).

11. Once the contents have drained through the well, apply a pulse of full pressure (10 psi for 30 s) to drive any residual methylene chloride into the collection wells.

12. Evaporate the eluent under a gentle stream of nitrogen until the wells are dry.

13. Reconstitute the samples in 500 µL of 0.1 % formic acid in water and methanol (90:10). Vortex the reconstituted extracts to ensure complete dissolution of the dried residue.

14. Analyze on LC-MS/MS.

***3.2 Instrument Operating Conditions***

1. Table 2 summarizes typical LC conditions.

2. Table 3 summarizes typical MS conditions.

3. Table 4 summarizes typical MRM conditions.

Each instrument should be individually optimized for best method performance.

**Table 2**
**Typical HPLC conditions**

| Cycle name | Analyst LC-Inj DLW Standard_Rev05 | | | |
|---|---|---|---|---|
| Wash solvent 1 | CLRW | | | |
| Wash solvent 2 | Methanol | | | |
| Injection volume | 2 µL | | | |
| Vacuum degassing | On | | | |
| Temperature | 30 °C | | | |
| A reservoir | 0.1 % HCOOH in CLRW | | | |
| B reservoir | 100 % Acetonitrile | | | |
| Gradient table | | | | |
| Step | Time (min) | Flow (µL/min) | A (%) | B (%) |
| 0 | 0 | 200 | 66 | 34 |
| 1 | 1 | 200 | 66 | 34 |
| 2 | 2 | 200 | 0 | 100 |
| 3 | 2.25 | 200 | 0 | 100 |
| 4 | 2.5 | 200 | 66 | 34 |
| 5 | 4 | 200 | 66 | 34 |

**Table 3**
**Typical mass spectrometer conditions**

| Parameter | Value |
|---|---|
| CUR | 20 |
| GS1 | 45 |
| GS2 | 40 |
| TEM | 700 |
| ihe | ON |
| CAD | 6 |
| IS | 5000 |
| EP | 5 |
| Scan type | MRM |
| Scheduled MRM | No |
| Polarity | Positive |
| Scan mode | N/A |
| Ion source | Turbo spray |
| Resolution Q1 | Unit |
| Resolution Q3 | Unit |
| Intensity Thres. | 0.00 cps |
| Settling time | 0.0000 ms |
| MR pause | 5.0070 ms |
| MCA | No |
| Step size | 0.00 Da |
| Dwell (ms) | 100 |

***3.3  Data Analysis***

1. Representative MRM chromatograms of carisoprodol and meprobamate in both urine and plasma are shown in Figs. 1 and 2, respectively.

2. The dynamic range for this assay is 100–10,000 ng/mL for each analyte. Samples exceeding this range can be diluted 5× or 10× as needed to achieve an accurate calculated concentration, if needed.

3. Data analysis is performed using the Analyst software to integrate peaks, calculate peak area ratios, and construct calibration curves using a linear 1/x weighted fit ignoring the origin

**Table 4**
**Typical MRM conditions**

| Analyte | Q1 Mass (Da) | | Q3 Mass (Da) | Param |
|---|---|---|---|---|
| Meprobamate (Quant.) | 219.1 | | 158.1 | DP = 35.0<br>CE = 15.0<br>CXP = 9.0 |
| | Approx. retention time | | 1.37 min | |
| Meprobamate (qualifier) | 219.1 | | 97.1 | DP = 35.0<br>CE = 15.0<br>CXP = 9.0 |
| | Approx. retention time | | 1.37 min | |
| Meprobamate-D7 (internal standard) | 226.1 | | 165.1 | DP = 35.0<br>CE = 15.0<br>CXP = 9.0 |
| | Approx. retention time | | 1.35 min | |
| Carisoprodol (Quant.) | 261.1 | | 176.1 | DP = 35.0<br>CE = 18.0<br>CXP = 9.0 |
| | Approx. retention time | | 2.87 min | |
| Carisoprodol (qualifier) | 261.1 | | 97.1 | DP = 35.0<br>CE = 18.0<br>CXP = 9.0 |
| | Approx. retention time | | 2.87 min | |
| Carisoprodol-D7 (internal standard) | 268.1 | | 183.1 | DP = 35.0<br>CE = 18.0<br>CXP = 9.0 |
| | Approx. retention time | | 2.85 min | |

as a data point. Sample concentrations are then calculated using the derived calibration curves (*see* **Note 1**).

4. Calibration curves should have an $r^2$ value $\geq 0.99$.

5. Typical imprecision is <15 % both inter- and intra-assay.

6. An analytical batch is considered acceptable if chromatography is acceptable and QC samples calculate to within 20 % of their target values and ion ratios are within 20 % of the calibration curve ion ratios.

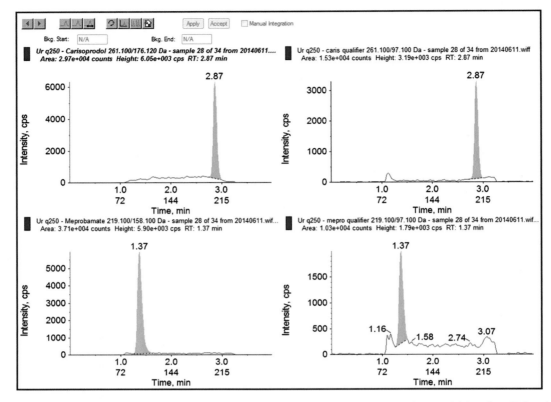

**Fig. 1** Typical MRM chromatogram for carisoprodol (2.87 min) and meprobamate (1.37 min) in urine. 250 ng/mL extracted from fortified human urine and analyzed according to the described method

## 4   Notes

1. Validate/verify all calibrators, QCs, internal standard, and negative matrix pools before placing into use.

2. Biotage SLE+ array wells are described in this method. Other equivalent products (96-well plates, columns) may be adapted as needed for use in this analysis.

3. This method describes a positive pressure extraction configuration. Vacuum manifolds may be adapted and configured for this method if needed.

4. A small pulse of positive pressure (2 s at 3 psi) can be applied to get the liquid to penetrate into the SLE bed, if necessary.

5. Allow all of the first methylene chloride aliquot to pass through the bed prior to adding the second aliquot.

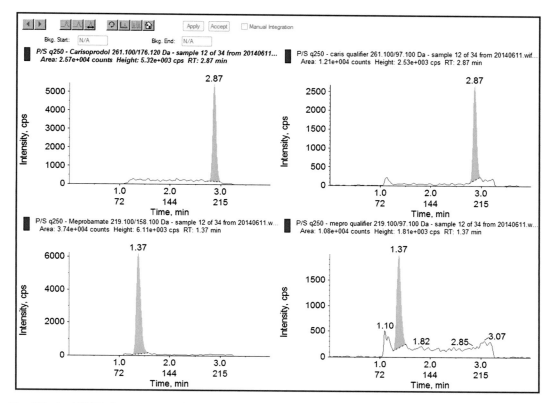

**Fig. 2** Typical MRM chromatogram for carisoprodol (2.87 min) and meprobamate (1.37 min) in plasma/serum. 250 ng/mL extracted from fortified human plasma and analyzed according to the described method

## References

1. Tse SA, Atayee RS, Ma JD, Best BM (2014) Factors affecting carisoprodol metabolism in pain patients using urinary excretion data. J Anal Toxicol 38(3):122–128. doi:10.1093/jat/bku002

2. McMillin GA, Slawson MH, Marin SJ, Johnson-Davis KL (2013) Demystifying analytical approaches for urine drug testing to evaluate medication adherence in chronic pain management. J Pain Palliat Care Pharmacother 27(4):322–339. doi:10.3109/15360288.2013.847889

3. Backer RC, Zumwalt R, McFeeley P, Veasey S, Wohlenberg N (1990) Carisoprodol concentrations from different anatomical sites: three overdose cases. J Anal Toxicol 14(5):332–334

4. Gaillard Y, Billault F, Pepin G (1997) Meprobamate overdosage: a continuing problem. Sensitive GC-MS quantitation after solid phase extraction in 19 fatal cases. Forensic Sci Int 86(3):173–180

5. Daval S, Richard D, Souweine B, Eschalier A, Coudore F (2006) A one-step and sensitive GC-MS assay for meprobamate determination in emergency situations. J Anal Toxicol 30(5):302–305

6. Baselt RC (2011) Disposition of toxic drugs and chemicals in man, 9th edn. Biomedical Publications, Seal Beach, CA

7. Alvarez JC, Duverneuil C, Zouaoui K, Abe E, Charlier P, de la Grandmaison GL, Grassin-Delyle S (2012) Evaluation of the first immunoassay for the semi-quantitative measurement of meprobamate in human whole blood or plasma using biochip array technology. Clin Chim Acta 413(1–2):273–277. doi:10.1016/j.cca.2011.10.025

8. Coulter C, Garnier M, Tuyay J, Orbita J Jr, Moore C (2012) Determination of carisoprodol and meprobamate in oral fluid. J Anal Toxicol 36(3):217–220. doi:10.1093/jat/bks009

9. Essler S, Bruns K, Frontz M, McCutcheon JR (2012) A rapid quantitative method of carisoprodol and meprobamate by liquid chromatography-tandem mass spectrometry. J Chromatogr B Analyt Technol Biomed Life Sci 908:155–160. doi:10.1016/j.jchromb.2012.09.001

10. Sreenivasulu V, Ramesh M, Kumar IJ, Babu RV, Pilli NR, Krishnaiah A (2013) Simultaneous determination of carisoprodol and aspirin in human plasma using liquid chromatography-tandem mass spectrometry in polarity switch mode: application to a human pharmacokinetic study. Biomed Chromatogr 27(2):179–185. doi:10.1002/bmc.2766

11. Bevalot F, Gustin MP, Cartiser N, Gaillard Y, Le Meur C, Fanton L, Guitton J, Malicier D (2013) Using bone marrow matrix to analyze meprobamate for forensic toxicological purposes. Int J Leg Med 127(5):915–921. doi:10.1007/s00414-013-0833-8

# Chapter 13

# Cetirizine Quantification by High-Performance Liquid Chromatography Tandem Mass Spectrometry (LC-MS/MS)

## Ada Munar, Clint Frazee, Bridgette Jones, and Uttam Garg

### Abstract

A multiple reaction monitoring (MRM), positive ion electrospray ionization, LC/MS/MS method is described for the quantification of cetirizine. The compound was isolated from human plasma by protein precipitation using acetonitrile. Cetirizine d4 was used as an internal standard. Chromatographic conditions were achieved using a C18 column and a combination of ammonium acetate, water, and methanol as the mobile phase. MRMs were: cetirizine, $389.26 \rightarrow 165.16, 201.09$; cetirizine d4, $393.09 \rightarrow 165.15, 201.10$. Calibration curves were constructed by plotting the peak area ratios of the calibrators' target MRM transition area to labeled internal standard target MRM transition area versus concentration.

**Key words** Cetirizine, Antihistamine, $H_1$-receptor, Allergic rhinitis and chronic urticaria

## 1 Introduction

Levocetirizine (RS)-2-[2-[4-[(2-Chlorophenyl)phenylmethyl] piperazin-1-yl]ethoxy]acetic Acid Dihydrochloride is a second-generation antihistamine and $R$-enantiomer of the cetirizine an active metabolite of hydroxyzine (a first-generation $H_1$-receptor inverse agonist) [1–4]. Cetirizine is a highly selective $H_1$-receptor inverse agonist and a potent non-sedating antihistamine [1–3]. As compared to other commonly used antihistamines, cetirizine has less affinity for calcium channel, adrenergic $\alpha_1$, dopamine $D_2$, serotonin $5\text{-}HT_2$ receptors, and muscarinic receptors. Cetirizine is minimally metabolized with an elimination half-life of approximately 8 h. The drug is 91 % protein bound, and also has a small volume of distribution (Vd) of 0.4 L/kg [5, 6]. Due to these pharmacologic properties, cetirizine is commonly prescribed to patients with allergic disease (e.g. allergic rhinitis and chronic urticaria) and is approved for use in children (2 years of age and older) and adults.

Various methods for cetirizine measurement including gas chromatography, high-performance liquid chromatography (HPLC) with UV or mass spectrometry detection have been

Uttam Garg (ed.), *Clinical Applications of Mass Spectrometry in Drug Analysis: Methods and Protocols*, Methods in Molecular Biology, vol. 1383, DOI 10.1007/978-1-4939-3252-8_13, © Springer Science+Business Media New York 2016

described in the literature [7–13]. We developed a simple, rapid, and highly sensitive method utilizing HPLC-tandem mass spectrometry for the measurement of cetirizine.

## 2  Materials

### 2.1  Samples

Heparinized plasma.

### 2.2  Solvents and Reagents

1. 7.5 M Ammonium acetate. Purchased as solution (Sigma-Aldrich, St Louis, MO).

2. Precipitating reagent containing internal standard: Add 200 μL of 50 μg/mL of secondary internal standard to a 100 mL volumetric flask and bring to volume with acetonitrile. Stable for 1 year at −20 °C.

3. Mobile phase A (20 mM ammonium acetate in water): To 1 L water, add 2.7 mL of 7.5 mM ammonium acetate and 570 μL of formic acid. Mix and degas. Store at ambient temperature. Stable for 1 month.

4. Mobile phase B (20 mM ammonium acetate in methanol): To 1 L methanol, add 2.7 mL of 7.5 mM ammonium acetate and 570 μL of formic acid. Mix and degas. Store at ambient temperature. Stable for 1 month.

5. Fresh frozen plasma: Obtain outdated fresh frozen plasma from blood bank or commercial source. Centrifuge at $2000 \times g$ for 10 min to remove particulates.

### 2.3  Internal Standards and Standards

1. Stock standard of Cetirizine (Sigma-Aldrich, St Louis, MO): Quantitatively prepare a 1 mg/mL stock standard of cetirizine, using cetirizine dihydrochloride, in methanol. Stable for 1 year when stored at −20 °C.

2. 100 μg/mL primary standard: Prepared by transferring 1 mL of stock standard to a 10 mL volumetric flask and diluting with methanol. Stable for 1 year when stored at −20 °C.

3. 10 μg/mL secondary standard: Prepared by transferring 1 mL of primary standard to a 10 mL volumetric flask and diluting with methanol. Stable for 1 year when stored at −20 °C.

4. 1 μg/mL tertiary standard: Prepare by transferring 1 mL of secondary standard to a 10 mL volumetric flask and diluting with methanol. Stable for 1 year when stored at −20 °C.

5. 1 mg/mL primary internal standard (Cetirizine-$d4$, C/D/N/Isotopes): Quantitatively prepare a 1 mg/mL primary standard of Cetirizine-$d4_I$ in methanol. Stable for 1 year when stored at −20 °C.

**Table 1**
**Preparation of calibrators using drug-free plasma**

| Calibrator | Primary standard (μL) | Secondary standard (μL) | Tertiary standard (μL) | Final concentration (ng/mL) |
|---|---|---|---|---|
| Blank | | | | |
| 1 | | | 10 | 1 |
| 2 | | 10 | | 10 |
| 3 | | 50 | | 50 |
| 4 | | 100 | | 100 |
| 5 | 50 | | | 500 |

The final volume of each calibrator is 10 mL

**Table 2**
**Preparation of quality controls using drug-free plasma**

| QC | Primary standard (μL) | Secondary standard (μL) | Tertiary standard (μL) | Final concentration (ng/mL) |
|---|---|---|---|---|
| 1 | | | 30 | 3 |
| 2 | 25 | | | 250 |
| 3 | 40 | | | 400 |

The final volume of each control is 10 mL

6. 50 μg/mL secondary internal standard: Prepare by transferring 500 μL of primary standard to a 10 mL volumetric flask and diluting with methanol. Stable for 1 year when stored at −20 °C.

**2.4 Calibrators and Controls**

1. Calibrators: Prepare calibrators 1–5 according to Table 1.

2. Quality Controls: Prepare control 1–3 according to Table 2.

For calibrators and controls add appropriate amount of standards to a 10 mL volumetric flask and qs to 10 mL with plasma (*see* **Note 1**).

**2.5 Analytical Equipment and Supplies**

1. AB Sciex LC-MS/MS 4000Q TRAP (Foster City, CA).

2. Shimadzu Prominence HPLC system with autosampler, two pumps and degasser (Lenexa, KS).

3. Autosampler vials with caps.

4. Analytical column: Supelcosil LC-18, 5 cm × 4.6 mm × 3 μm (Sigma-Aldrich, St Louis, MO).

5. Guard column: Pinnacle, C18, 10 mm × 4 mm × 5 μm (Restek, Belfonte, PA).

## 3    Methods

### 3.1    Stepwise Procedure

1. Pipette 100 μL of well mixed calibrators, patient plasma and controls to the appropriately labeled microcentrifuge tubes.
2. Add 200 μL of precipitating reagent containing internal standards to each tube.
3. Immediately cap the samples and vortex for ~20 s.
4. Rock the tubes for 10 min.
5. Centrifuge the tubes at $10,000 \times g$ for 5 min.
6. Using disposable tips transfer 200 μL of supernatant into autosampler (*see* **Note 2**).
7. Inject 10 μL into liquid chromatography-tandem mass spectrometry (LC-MSMS) for analysis.

### 3.2    Instrument's Operating Conditions

Instrument's operating conditions are given in Table 3.

### 3.3    Data Analysis

1. Data are collected and analyzed using Analyst 1.5.1 software (AB Sciex, Foster City, CA).
2. Calibration curves are constructed from peak area ratios of MRM of calibrators and internal standards versus concentration.

**Table 3**
**Instrument's operating conditions**

| A. HPLC | | |
|---|---|---|
| Time (min) | Mobile phase A (%) | Mobile phase B (%) |
| 0.5 | 50 | 50 |
| 1.5 | 0 | 100 |
| 3 | 0 | 100 |
| 3.1 | 50 | 50 |
| **B. MS/MS parameters** | | |
| Source (electrospray ionization, positive mode) | | |
| Curtain gas | 25 psi | |
| Source temperature | 375 °C | |
| Collision gas (CAD) | High | |
| Ion source gas 1 (GS1) | 50 psi | |
| Ion source gas 1 (GS2) | 60 psi | |

Column temperature—65 °C, Flow rate—1.0 mL/min

**Table 4**
**Multiple reaction monitoring transitions**

| Analyte | Q1 mass (amu) | Q3 mass (amu) | Qualifier ion |
|---|---|---|---|
| Cetirizine | 389.26 | 165.16 | 201.09 |
| Cetirizine-d4 | 393.09 | 165.15 | 201.10 |

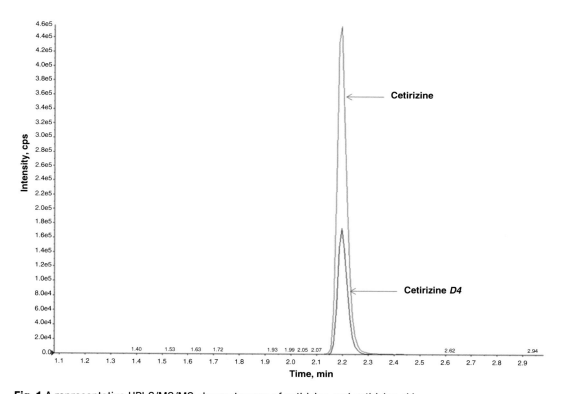

**Fig. 1** A representative HPLC/MS/MS chromatogram of cetirizine and cetirizine-d4

3. A typical calibration curve has a correlation ($r^2$) >0.99.

4. Multiple reaction monitoring transitions for each analyte are given in Table 4.

5. A typical HPLC/MS/MS chromatogram of cetirizine is shown in Fig. 1.

6. Quality control samples are evaluated with each run. The run is considered acceptable if calculated concentrations of controls are within the ±20 % of target values.

7. Samples with results greater than upper limit of linearity should be diluted with blank.

## 4 Notes

1. When possible, calibrators and controls should be prepared from different lot of stock solution on separate days.

2. Be sure not to disturb the pellet.

### References

1. Bachert C (2005) Levocetirizine: a modern H1-antihistamine for the treatment of allergic rhinitis. Expert Rev Clin Immunol 1:495–510

2. Kapp A, Wedi B (2004) Chronic urticaria: clinical aspects and focus on a new antihistamine, levocetirizine. J Drugs Dermatol 3:632–639

3. Kranke B, Mayr-Kanhauser S (2005) Urticarial reaction to the antihistamine levocetirizine dihydrochloride. Dermatology 210:246–247

4. Walsh GM (2008) A review of the role of levocetirizine as an effective therapy for allergic disease. Expert Opin Pharmacother 9:859–867

5. Passalacqua G, Canonica GW (2005) A review of the evidence from comparative studies of levocetirizine and desloratadine for the symptoms of allergic rhinitis. Clin Ther 27:979–992

6. Morita MR, Berton D, Boldin R, Barros FA, Meurer EC, Amarante AR, Campos DR, Calafatti SA, Pereira R, Abib E Jr et al (2008) Determination of levocetirizine in human plasma by liquid chromatography-electrospray tandem mass spectrometry: application to a bioequivalence study. J Chromatogr B Analyt Technol Biomed Life Sci 862:132–139

7. Rosseel MT, Lefebvre RA (1991) Determination of cetirizine in human urine by high-performance liquid chromatography. J Chromatogr 565:504–510

8. Ren XL, Tian Y, Zhang ZJ, Chen Y, Wu LL, Huang J (2011) Determination of cetirizine in human plasma using high performance liquid chromatography coupled with tandem mass spectrometric detection: application to a bioequivalence study. Arzneimittelforschung 61:287–295

9. Pandya KK, Bangaru RA, Gandhi TP, Modi IA, Modi RI, Chakravarthy BK (1996) High-performance thin-layer chromatography for the determination of cetirizine in human plasma and its use in pharmacokinetic studies. J Pharm Pharmacol 48:510–513

10. Moncrieff J (1992) Determination of cetirizine in serum using reversed-phase high-performance liquid chromatography with ultraviolet spectrophotometric detection. J Chromatogr 583:128–130

11. Macek J, Ptacek P, Klima J (1999) Determination of cetirizine in human plasma by high-performance liquid chromatography. J Chromatogr B Biomed Sci Appl 736:231–235

12. Kang SW, Jang HJ, Moore VS, Park JY, Kim KA, Youm JR, Han SB (2010) Enantioselective determination of cetirizine in human plasma by normal-phase liquid chromatography-atmospheric pressure chemical ionization-tandem mass spectrometry. J Chromatogr B Analyt Technol Biomed Life Sci 878:3351–3357

13. de Jager AD, Hundt HK, Swart KJ, Hundt AF, Els J (2002) Extractionless and sensitive method for high-throughput quantitation of cetirizine in human plasma samples by liquid chromatography-tandem mass spectrometry. J Chromatogr B Analyt Technol Biomed Life Sci 773:113–118

# Chapter 14

## Quantification of Docetaxel in Serum Using Turbulent Flow Liquid Chromatography Electrospray Tandem Mass Spectrometry (TFC-HPLC-ESI-MS/MS)

### Christopher A. Crutchfield, Mark A. Marzinke, and William A. Clarke

### Abstract

Docetaxel is a second-generation taxane and is used clinically as an anti-neoplastic agent in cancer chemotherapy via an anti-mitotic mechanism. Its efficacy is limited to a narrow therapeutic window. Inappropriately high concentrations may cause erythema, fluid retention, nausea, diarrhea, and neutropenia. As a result, dosing recommendations have changed from high dosage loading every 3 weeks to lower dosage loading weekly. We describe a method that can be used for therapeutic drug monitoring of docetaxel levels using turbulent flow liquid chromatography electrospray tandem mass spectrometry (TFC-HPLC-ESI-MS/MS). The method is rapid, requiring only 6.3 min per analytical run following a simple protein crash. The method requires only 100 μL of serum. Concentrations of docetaxel were quantified by a calibration curve relating the peak-area ratio of docetaxel to a deuterated internal standard (docetaxel-D9). The method was linear from 7.8 to 1000 ng/mL, with imprecision ≤6.2 %.

**Key words** Docetaxel, Anti-neoplastic, Turbulent flow liquid chromatography electrospray tandem mass spectrometry

## 1 Introduction

Docetaxel is a chemotherapeutic used to treat solid tumors, including breast, non-small-cell lung, and prostate cancer [1–3]. However, its use can cause side effects including erythema, nausea, diarrhea, fluid retention, and neutropenia [1]. Initially docetaxel had a standard dosage of 75–100 mg/m² over 1 h every 3 weeks [1], however later studies suggested a lower dose of 20–40 mg/m² every 1 week [4]. Docetaxel is highly protein bound [5] and its pharmacological effect is tied to its free form. With frequent dosing and high protein affinity, clinical teams may find utility in monitoring the pharmacokinetics of docetaxel in their patient population. This method [6] enables rapid analysis of docetaxel using turboflow on-line extraction prior to analytical separation and analysis using tandem mass spectrometry.

Uttam Garg (ed.), *Clinical Applications of Mass Spectrometry in Drug Analysis: Methods and Protocols*, Methods in Molecular Biology, vol. 1383, DOI 10.1007/978-1-4939-3252-8_14, © Springer Science+Business Media New York 2016

## 2    Materials

### 2.1    Sample

Human serum.

### 2.2    Solvents and Reagents

1. Mobile Phase A, 0.1 % (v/v) formic acid in HPLC-grade water, stable for 1 month at room temperature, 18–24 °C.

2. Mobile Phase B, 0.1 % (v/v) formic acid in HPLC-grade methanol, stable for 1 month at room temperature, 18–24 °C.

3. Mobile Phase C, 40:40:20 acetonitrile:isopropanol:acetone.

4. Human drug-free normal serum.

### 2.3    Standards and Internal Standard

1. Primary standard of docetaxel was prepared by dissolving docetaxel powder (Toronto Research Chemicals) in methanol at a final concentration of 5 mg/mL.

2. Primary internal standard was prepared by dissolving docetaxel-d9 powder (Toronto Research Chemicals) in methanol at a final concentration of 1 mg/mL.

3. Primary working solutions are prepared by diluting the primary standard solution to concentrations of 100, 10, and 1 µg/mL.

4. I.S. Working Solution/Extraction Solution (500 ng/mL docetaxel-d9) was prepared by adding 500 µL of the 1 mg/mL primary internal standard to a Class A 1000 mL volumetric flask, filling to the level, and mixing.

### 2.4    Calibrators and Controls

1. Calibrators: Prepare calibrators 1–9 by diluting working stock solutions with drug-free normal human serum in 10 mL class A volumetric flasks (Table 1).

2. Controls: Using independently prepared working stock solutions, prepare low, medium, and high QC levels at 50, 250, and 1000 ng/mL.

### 2.5    Analytical Equipment and Supplies

1. Aria TLX1 system equipped with a CTC HTC PAL Autosampler and two Agilent 1250 Pumps coupled to a Thermo TSQ Vantage triple quadrupole mass spectrometer (Thermo Fisher Scientific).

2. Pre-Analytical column: Thermo Cyclone-P 0.5×50 mm (Thermo Fisher Scientific).

3. Analytical column: Thermo Scientific Hypersil Gold C-18 2.1×50 mm, particle size 3 µm.

4. 1.8 mL glass HPLC vials.

5. 1.5 polypropylene microcentrifuge tubes.

**Table 1**
**Preparation of calibrators**

| Calibrator | Working stock concentration (µg/mL) | Working stock volume (µL) | Final volume (mL) | Final concentration (ng/mL) |
|---|---|---|---|---|
| 1 | 1 | 78 | 10 | 7.8 |
| 2 | 1 | 156 | 10 | 15.6 |
| 3 | 1 | 313 | 10 | 31.3 |
| 4 | 1 | 625 | 10 | 62.5 |
| 5 | 10 | 125 | 10 | 125 |
| 6 | 10 | 250 | 10 | 250 |
| 7 | 10 | 500 | 10 | 500 |
| 8 | 100 | 100 | 10 | 1000 |
| 9 | 100 | 200 | 10 | 2000 |

## 3    Methods

### 3.1    Stepwise Procedure

1. To a labeled 1.5 mL polypropylene centrifuge tube, pipette 100 µL of serum (calibrator, control, or unknown sample) (*see* **Note 1**).
2. Add 300 µL of extraction solution.
3. Cap and vortex for 20 s.
4. Centrifuge for 5 min at $18,000 \times g$.
5. Dilute 300 µL supernatant 1:1 with HPLC-grade water in a labeled 1.8 mL glass vial.
6. Cap and vortex briefly.
7. Please vials into autosampler.
8. Inject 25 µL and analyze.

### 3.2    Sample Analysis

1. Instrumental operating parameters are given in Table 2.
2. Data are analyzed using LCQuan (Thermo Scientific).
3. Standard curves are generated based on linear regression with $1/x^2$ weighting of the analyte/internal standard peak-area ratio relative to the nominal analyte concentration.
4. Imprecision is typically ≤6.2 % at all QC levels.

**Table 2**
**HPLC-MS/MS operation conditions**

| A. HPLC | | | | | | | | | | |
|---|---|---|---|---|---|---|---|---|---|---|
| Time (min) | Length (s) | TX flow rate (mL/min) | TX mobile phase A (%) | TX mobile phase B (%) | TX mobile phase C (%) | Tee | Loop | LX flow rate (mL/min) | LX mobile phase A (%) | LX mobile phase B (%) |
| 0 | 30 | 1.50 | 100 | 0 | 0 | | Out | 0.25 | 100 | 0 |
| 0.5 | 45 | 0.20 | 25 | 75 | 0 | Tee | In | 0.50 | 100 | 0 |
| 1.25 | 200 | 1.50 | 0 | 0 | 100 | | In | 0.25 | 10 | 90 |
| 4.58 | 45 | 1.50 | 25 | 0 | 100 | | In | 0.25 | 0 | 100 |
| 5.33 | 60 | 1.50 | 100 | 0 | 0 | | Out | 0.25 | 100 | 0 |

| B. MS/MS tune settings | |
|---|---|
| Parameter | Value |
| Spray voltage (V) | 3500 |
| Sheath gas | 35 |
| Aux gas | 35 |
| Capillary temperature (°C) | 200 |

| C. Precursor and product ions | | | |
|---|---|---|---|
| Compound | Precursor | Product | CE (eV) |
| Docetaxel | 808.4 | 225.9 | 10 |
| Docetaxel-D9 | 817.4 | 226.9 | 10 |

# 4 Notes

1. Matrix effects were evaluated using post-column infusion as well as comparison of spiked sera and spiked solvent. Matrix effects were <14 %.

## References

1. Salminen E, Bergman M, Huhtala S, Ekholm E (1999) Docetaxel: standard recommended dose of 100 mg/m(2) is effective but not feasible for some metastatic breast cancer patients heavily pretreated with chemotherapy-A phase II single-center study. J Clin Oncol 17:1127

2. Saloustros E, Georgoulias V (2008) Docetaxel in the treatment of advanced non-small-cell lung cancer. Expert Rev Anticancer Ther 8:1207–1222

3. Gilbert DC, Parker C (2005) Docetaxel for the treatment of prostate cancer. Future Oncol 1:307–314

4. Burstein HJ, Manola J, Younger J et al (2000) Docetaxel administered on a weekly basis for metastatic breast cancer. J Clin Oncol 18:1212–1219

5. Clarke DSJ, Rivory LP (2012) Clinical pharmacokinetics of docetaxel. Clin Pharmacokinet 36:99–114

6. Marzinke MA, Breaud AR, Clarke W (2013) The development and clinical validation of a turbulent-flow liquid chromatography-tandem mass spectrometric method for the rapid quantitation of docetaxel in serum. Clin Chim Acta 417:12–18

# Chapter 15

# Comprehensive Urine Drug Screen by Gas Chromatography/Mass Spectrometry (GC/MS)

## Bheemraj Ramoo, Melissa Funke, Clint Frazee, and Uttam Garg

## Abstract

Drug screening is an essential component of clinical toxicology laboratory service. Some laboratories use only automated chemistry analyzers for limited screening of drugs of abuse and few other drugs. Other laboratories use a combination of various techniques such as immunoassays, colorimetric tests, and mass spectrometry to provide more detailed comprehensive drug screening. Mass spectrometry, gas or liquid, can screen for hundreds of drugs and is often considered the gold standard for comprehensive drug screening. We describe an efficient and rapid gas chromatography/mass spectrometry (GC/MS) method for comprehensive drug screening in urine which utilizes a liquid–liquid extraction, sample concentration, and analysis by GC/MS.

**Key words** Toxicology, GC/MS, Drug analysis, Comprehensive drug screening, Drug screening

## 1 Introduction

Comprehensive drug screening is a necessity for full-service clinical toxicology laboratories and may utilize a combination of various techniques such as colorimetric spot tests, immunoassays, gas chromatography, and mass spectrometry [1–4]. Colorimetric spot tests are neither sensitive nor specific. Immunoassays are the most commonly used method for drug screening in most clinical laboratories as they are simple and available on most automated chemistry analyzers. However, immunoassays need special antibody reagents, and are not very specific. Gas chromatography linked to flame ionization detector is used for volatile (ethanol, methanol, isopropanol, and acetone) analysis. Mass spectrometry is used for broad spectrum drug screening [2, 4, 5]. It is one of the most sensitive and specific methods available for drug detection. The other major advantage of mass spectrometry is that it does not require special reagents. This is important in light of the ever-growing availability of clandestine synthetic drugs.

Uttam Garg (ed.), *Clinical Applications of Mass Spectrometry in Drug Analysis: Methods and Protocols*, Methods in Molecular Biology, vol. 1383, DOI 10.1007/978-1-4939-3252-8_15, © Springer Science+Business Media New York 2016

Gas or liquid chromatography-mass spectrometry can be used to screen 100–1000 s of drugs [2, 5–8]. Gas chromatography-mass spectrometry has been the reference method for many years, and still is the most widely used method for drug screening. Advantages of GC/MS are low cost and the universal availability of several commercial drug identification libraries [9]. Primary disadvantage of GC/MS is that detection is limited to drug volatility. Due to this disadvantage, many laboratories now use liquid chromatography-tandem mass spectrometry (LC/MS/MS) for drug screening [7, 8, 10, 11]. The major disadvantages of LC/MS/MS, however, are higher costs and lack of universal commercial drug libraries. In our laboratory, we use GC/MS method described here for comprehensive drug screening. Drugs from urine are extracted using a liquid/liquid extraction technique. The extract is concentrated, reconstituted, and analyzed by GC/MS using full-scan electron impact ionization mode. The method is validated for detecting greater than 150 drugs within a multitude of drug classes. Drug spectra are indentified through electronic library searches using commercial libraries. In addition, drug identification is verified by relative retention time (RRT) through comparison of the sample drug's RRTs to pre-established in-house RRTs of corresponding drug reference standards. Though this GC/MS methodology provides very specific drug identification criteria, all drugs are reported with a comment advising confirmation of drug screen results if clinically indicated.

## 2  Materials

### 2.1  Samples

2–3 mL urine (*see* **Note 1**).

### 2.2  Solvents and Reagents

1. Promazine, ACS grade.
2. Methylene Chloride, High-Resolution GC grade.
3. Buffer Salts Mixture:
   (a) Using a 1 L graduated cylinder, measure 600 mL volume of sodium chloride.
   (b) Add 100 mL volume of sodium carbonate.
   (c) Add 100 mL volume of sodium bicarbonate.
   (d) Pour this mixture into a large amber jar and mix well by inverting the jar several times.
   (e) Add an additional 600 mL volume of sodium chloride to the amber jar and mix well by inverting the jar several times.
   (f) Stable at room temperature for 2 years.

4. Extraction Solvent:

   (a) Add 450 mL of methylene chloride to a 1 L amber oxford dispenser.

   (b) Add 450 mL of cyclohexanes.

   (c) Add 100 mL of isopropanol.

   (d) Add 400 μL of 1.7 mg/mL promazine. Mix well.

   (e) Stable at room temperature for 1 year.

5. Extraction Tubes:

   (a) Add approximately 1 g of buffer salts mixture to each 16×100 screw top extraction tube.

   (b) Add 3 mL extraction solvent.

   (c) Cap with Teflon lined caps only and store at room temperature until use.

   (d) Stable at room temperature for 1 year.

**2.3 Internal Standard and Controls**

1. Internal Standard, promazine, 1.7 mg/mL: Add 19.18 mg promazine-HCl stock powder to a 10 mL volumetric flask and qs to volume with methanol. Stable 1 year at room temperature.

2. Negative Control: Certified negative urine (Utak Laboratories) (*see* Fig. 1).

3. Positive Control:

   (a) *See* Table 1. Add the indicated volumes of certified Cerilliant drug standards as shown in Table 1 to a 100 mL volumetric flask and qs to 100 mL with negative urine (Utak Laboratories).

   (b) Evaluate in-house positive urine control at least daily by GC-MS for the presence of all expected drugs (*see* Fig. 2).

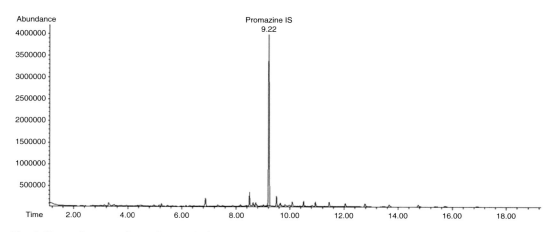

**Fig. 1** Chromatogram of negative control

**Table 1**
**Preparation of positive urine control**

| Drug/drug concentration | Volume (μL) | Final concentration (μg/mL) |
|---|---|---|
| Amphetamine 1 mg/mL | 300 | 3 |
| Ecgonine methyl ester 1 mg/mL | 100 | 1 |
| Imipramine 1 mg/mL | 300 | 3 |
| Meperidine 1 mg/mL | 200 | 2 |
| Methadone 1 mg/mL | 100 | 1 |
| Morphine 1 mg/mL | 300 | 3 |
| Oxycodone 1 mg/mL | 150 | 1.5 |
| Phenobarbital 1 mg/mL | 500 | 5 |
| Propoxyphene 1 mg/mL | 400 | 4 |
| Secobarbital 1 mg/mL | 100 | 1 |

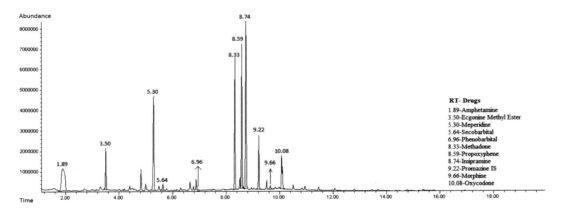

**Fig. 2** Chromatogram of positive control

**2.4 Analytical Equipment and Supplies**

1. Zymark Turbovap nitrogen evaporator.

2. Agilent GC-MS 5975C inert XL MSD with Triple Axis Detector (Agilent Technologies, CA). Perform autotune daily.

3. Analytical column: ZB-1MS 15 m×250 μm×0.25 μm (Phenomenex, Torrance, CA).

4. Carrier Gas: Helium, Ultrahigh Purity Grade.

5. 16×100 mm screw-cap test tubes with Teflon caps.

6. Autosampler vials with glass inserts and crimp caps (P.J. Cobert Associates, Inc., St. Louis, MO).

# 3 Methods

### 3.1 Stepwise Procedure

1. Aliquot 1 mL urine to already prepared extraction tube.
2. Rock or rotate tube for a minimum of 5 min.
3. Centrifuge at $2000 \times g$ for 5 min.
4. Transfer top organic layer to concentration tube and evaporate to dryness in Turbovap nitrogen evaporator at 40 °C using a gentle stream of nitrogen. Be careful not to over-dry (Two drops of acidified methanol, 5 %, may be added to extract prior to evaporation, *see* **Note 2**).
5. Reconstitute the dried extract with 125 μL Hexanes:Ethanol (1:1) and transfer to a labeled autosampler vial containing a flanged insert.
6. Analyze by GC-MS using the instrument conditions indicated in Table 2.

### 3.2 Instrument's Operating Conditions

*See* Table 2.

### 3.3 Data Analysis

1. Each chromatographic peak total ion chromatogram (TIC) should be carefully evaluated using electronic commercial and in-house libraries (e.g. AAFS, SOFT, TIAFT, Wiley6N, PMW_TOXR, etc.). If necessary, background subtraction is applied

**Table 2**
**GC-MS operating conditions**

| | |
|---|---|
| Initial oven temp. | 90 °C |
| Initial time | 1.0 min |
| Ramp 1 | 32 °C/min |
| Temp. 2 | 170 °C |
| Hold time | 2.0 min |
| Ramp 2 | 20 °C/min |
| Final temp. | 270 °C |
| Hold time | 9.5 min |
| Injector temp. | 250 °C |
| Purge time on | 1 min |
| Column pressure | 5 psi |
| Detector temp. | 280 °C |
| Detector mode | Electron impact at 70 eV |
| Tune | Autotune |

when interfering $m/z$ ratios are indicated. Always proceed cautiously when identifying very small peaks relative to the baseline and whenever case history and other findings are inconsistent.

2. GC/MS drug identifications are evaluated for relative retention time (RRT) acceptability. RRT is calculated by dividing the drug retention time by the internal standard retention time. We recommend an RRT agreement within ±0.01 of pre-established reference standards and a high-quality reference spectral match. Because RRT values can shift, reference RRT data should be re-established as necessary.

3. All positive results should be considered presumptive in the absence of confirmation by a second methodology using an independent chemical principle.

4. *See* **Notes 3–6** during data interpretation.

## 4   Notes

1. Cloudy or turbid urine should be centrifuged prior to analysis.

2. Acidified methanol is used to prevent any highly volatile drugs such as amphetamines, from evaporating during the drying process.

3. During GC-MS analysis, a sample containing a high level of diphenhydramine may result in multiple modafinil-like peaks (e.g. modafinil, modafinil artifact, modafinil breakdown) when evaluated by certain commercial libraries. This may be a result of diphenhydramine pyrolysis during GC-MS analysis.

4. Oxcarbazepine may produce false-positive carbamazepine results using this methodology.

5. Methocarbamol may produce false-positive guaifenesin results using this methodology due to thermal degradation in the injection port.

6. Dextromethorphan and levomethorphan are indistinguishable using this methodology.

## References

1. Garg U, Frazee CC, Scott D, Wasserman G (2004) Comprehensive toxicology drug screening data in a pediatric population. J Toxicol Clin Toxicol 42:681–683

2. Greller HA, Barrueto F Jr (2004) Comprehensive drug screening. Emerg Med J 21:646

3. Griffiths WC, Oleksyk SK, Diamond I (1973) A comprehensive gas chromatographic drug screening procedure for the clinical laboratory. Clin Biochem 6:124–131

4. Valli A, Polettini A, Papa P, Montagna M (2001) Comprehensive drug screening by integrated use of gas chromatography/mass spectrometry and Remedi HS. Ther Drug Monit 23:287–294

5. Maurer HH (2006) Hyphenated mass spectrometric techniques-indispensable tools in clinical

and forensic toxicology and in doping control. J Mass Spectrom 41:1399–1413

6. Fabbri A, Marchesini G, Morselli-Labate AM, Ruggeri S, Fallani M, Melandri R, Bua V, Pasquale A, Vandelli A (2003) Comprehensive drug screening in decision making of patients attending the emergency department for suspected drug overdose. Emerg Med J 20:25–28

7. Kamel A, Prakash C (2006) High performance liquid chromatography/atmospheric pressure ionization/tandem mass spectrometry (HPLC/API/MS/MS) in drug metabolism and toxicology. Curr Drug Metab 7:837–852

8. Peters FT (2011) Recent advances of liquid chromatography-(tandem) mass spectrometry in clinical and forensic toxicology. Clin Biochem 44:54–65

9. Aebi B, Bernhard W (2002) Advances in the use of mass spectral libraries for forensic toxicology. J Anal Toxicol 26:149–156

10. Lee YW (2013) Simultaneous screening of 177 drugs of abuse in urine using ultra-performance liquid chromatography with tandem mass spectrometry in drug-intoxicated patients. Clin Psychopharmacol Neurosci 11:158–164

11. Remane D, Wetzel D, Peters FT (2014) Development and validation of a liquid chromatography-tandem mass spectrometry (LC-MS/MS) procedure for screening of urine specimens for 100 analytes relevant in drug-facilitated crime (DFC). Anal Bioanal Chem 406:4411–4424

# Chapter 16

# Broad-Spectrum Drug Screening Using Liquid Chromatography-Hybrid Triple-Quadrupole Linear Ion Trap Mass Spectrometry

## Judy Stone

### Abstract

Urine is processed with a simple C18 solid-phase extraction (SPE) and reconstituted in mobile phase. The liquid chromatography system (LC) injects 10 μL of extracted sample onto a reverse-phase LC column for gradient analysis with ammonium formate/acetonitrile mobile phases. Drugs in the column eluent become charged in the ion source using positive electrospray ionization (ESI). Pseudomolecular ions (M + H) are analyzed by a hybrid triple-quadrupole linear ion trap (QqQ and QqLIT) mass spectrometer using an SRM-IDA-EPI acquisition. An initial 125 compound selected ion monitoring (SRM) survey scan (triple quadrupole or QqQ mode) is processed by the information-dependent acquisition (IDA) algorithm. The IDA algorithm selects SRM signals from the survey scan with a peak height above the threshold (the three most abundant SRM signals above 1000 cps) to define precursor ions for subsequent dependent scanning. In the dependent QqLIT scan(s), selected precursor ion(s) are passed through the first quadrupole (Q1), fragmented with three different collision energies in the collision cell (Q2 or q), and product ions are collected in the third quadrupole (Q3), now operating as a linear ion trap (LIT). The ions are scanned out of the LIT in a mass dependent manner to produce a full-scan product ion spectrum ($m/z$ 50–700) defined as an Enhanced (meaning acquired in LIT mode) Product Ion (EPI) spectrum (Mueller et al., Rapid Commun Mass Spectrom 19:1332–1338, 2005). Each EPI spectrum is linked to its precursor ion and to the associated SRM peak from the survey scan. EPI spectra are automatically searched against a 125 drug library of reference EPI spectra for identification. When the duty cycle is complete (one survey scan of 125 SRMs plus 0–3 dependent IDA-EPI scans) the mass spectrometer begins another survey scan of the 125 SRMs.

**Key words** Broad-spectrum drug screening, Urine solid-phase extraction, Positive-mode electrospray, Hybrid triple-quadrupole linear ion trap, Liquid chromatography-tandem mass spectrometry (LC-MSMS)

## 1 Introduction

The toxicologic process of testing body fluids for the widest possible range of drugs and poisons has been described as Comprehensive Drug Screening (CDS), General Unknown Screening (GUS), and Systematic Toxicological Analysis (STA) [1]. The unrealized expectation of GUS is that the methods used

Uttam Garg (ed.), *Clinical Applications of Mass Spectrometry in Drug Analysis: Methods and Protocols*, Methods in Molecular Biology, vol. 1383, DOI 10.1007/978-1-4939-3252-8_16, © Springer Science+Business Media New York 2016

have the capability to detect all possible chemicals of interest [1]. Different chromatographic-mass spectrometry techniques fulfill this goal to a greater or lesser degree based on sample preparation, chromatographic, ionization, acquisition mode, and mass analyzer characteristics [2–4]. Historically the use of two orthogonal methods—for example gas chromatography-mass spectrometry (GC-MS) and liquid chromatography with an ultraviolet detector (LC-UV) has identified the broadest possible menu of drugs [3, 4]. GCMS has higher resolution, greater specificity, and lower detection limits than does LC-UV but requires hydrolysis and derivatization steps in order to detect polar and/or heat-labile compounds, such as glucuronide/sulfate metabolites [3, 4].

The advent of ultrahigh-performance liquid chromatography/ atmospheric pressure ionization (API) coupled to tandem mass spectrometry appeared to offer the potential for optimal GUS using a single method [1, 2]. The very broad range of compounds that can be ionized with electrospray ionization sources (ESI) means that more compounds of interest can be detected with a single technique. However, ionization of endogenous compounds and environmental contaminants is also enhanced. Even with selective sample preparation, a highly complex mélange of ions occurs when biological samples such as urine or serum undergo ESI [1, 2]. It has proved challenging to tease out all ions that represent compounds of interest from the high background noise [5]. Although sample preparation techniques such as SPE or liquid-liquid extraction can successfully reduce the interference from matrix compounds, potential drugs or toxins of interest may also be lost during sample preparation. The enhanced sensitivity and selectivity of tandem (dual stage) mass analysis using a "multi-targeted" data-dependent or information-dependent acquisition strategy (DDA or IDA) as described in this method is one approach for dealing with these issues [6]. A multi-targeted DDA method looks for only those compounds defined in the acquisition—e.g., the 125 SRM transitions referred to in the Summary—and ignores all other ions—e.g., matrix and environmental background noise [6]. One disadvantage of a multi-targeted approach is obvious—any drug not included in the target list will not be detected. A true unknown screening approach is non-targeted, using a data-independent acquisition (DIA) to look for any compound that has been extracted, introduced by chromatography, and ionized [1].

The degrees to which broad-spectrum drug detection is improved more by the choice of sample preparation method versus the use of dual-stage or single-stage mass analysis versus the use of data-dependent acquisition (DDA) or data-independent acquisition (DIA) are difficult to determine, although one study comparing LC-UV, GCMS, LC-MS, and two LC-MSMS methods suggests that dual-stage mass analysis may be the instrument architecture of choice [5]. Recent refinements in tandem high-resolution mass analysis using a novel DIA strategy (Sequential Windowed Acquisition of All Theoretical Fragment Ion Mass Spectra or SWATH) appear promising, potentially combining the greater selectivity of DDA with the more comprehensive detection options of IDA [7, 8].

Decisions about drug screening techniques should take into account the clinical reason for performing drug testing. When a greater degree of confidence is required that a selected list of drugs will be correctly determined as present or not present, e.g., when monitoring for compliance, a multi-targeted, DDA strategy may be more reliable. When a search for any and all toxins as a cause of an unexplained pathology is appropriate—a non-targeted, DIA approach, although more time consuming for data analysis, may identify additional compounds not found by a more selective method.

# 2  Materials

## 2.1  Reagents and Buffers (See Note 1)

1. Ammonium formate buffer 1 mol/L, LC-MS grade (for Mobile Phase A).

2. Human drug-free urine.

3. 1.0 N Potassium hydroxide (1.0 N KOH) for pH adjustment of mobile phase A.

4. 0.1 M Phosphate buffer, pH 6.0 (SPE application buffer).

5. Mobile phase reagent A: 2 mmol/L ammonium formate (pH 3).

6. Mobile phase reagent B: 10:90 (vol:vol) mobile phase A:acetonitrile.

7. SPE elution solvent: 73:25:2 (vol:vol) ethyl acetate: isopropanol:$NH_4OH$.

8. Reconstitution (injection) solvent: 90:10 (vol:vol) mobile phase A:mobile phase B.

## 2.2  Library Reference Standards (Calibrators/ Calibration)

1. This is a qualitative method; no calibrators were used and no concentrations were reported.

2. Instead of a lower limit of quantitation, the criteria necessary for reporting a drug as "presumptive positive" were (a) peak area must be $\geq 5e + 3$; (b) purity (library match factor for EPI spectra) must be $\geq 70$ and the spectral match must pass visual review criteria; (c) relative retention time (RRt) must be within $\pm 0.05$ min or 5 %, whichever is larger, of the library reference standard RRt; (d) the drug must not be present in the negative control (drug free urine); and (e) both internal standards must be identified in the sample with peak areas, spectral purity, and Rts within acceptable limits.

3. Reference standard RRts and retention times (Rts) were established by spiking drug standards, five standards per sample, into drug-free urine at a concentration of 500 µg/L followed by analysis using the procedure described.

4. From the same analyses of drug standards as in **item 3**, acquired EPI spectra from drug standards were searched against existing AB Sciex libraries and compared to reference spectra available online, reviewed for acceptability, and used to create an in-house spectral library containing only the 125 compounds in the method. We created the library in-house (yielding more similar reference and patient spectra) and added only those compounds that were targeted in the SRM survey scan (resulting in fewer irrelevant library hits) in order to decrease the time required to exclude false-positive hits from automated library searches and uncover false negatives with manual library search.

5. Primary drug standards in methanol, 1.0 and 0.1 mg/mL, or powders (when standards in methanol were unavailable) were purchased to create standards used for the spectral library.

**2.3  Internal Standard and Quality Controls**

1. The working internal standard solution contained hydromorphone-D6 and diazepam-D5 in acetonitrile at a concentration of 500 µg/L (*see* **Note 2**).

2. Quality control samples: Drug-free urine (blank or negative control) and a commercial lyophilized urine broad-spectrum drug screen control, reconstituted with water and then diluted ×10 (1:9) with drug-free urine, were used to validate performance daily. The 11 drugs in the positive control, their concentrations after ×10 dilution, and typical Rt, RRt, and peak areas are shown in Table 1.

**2.4  Supplies**

1. 13 × 100 mm borosilicate glass test tubes.

2. Autosampler vials and screw caps with PTFE-lined septa (2 mL amber, wide opening, borosilicate glass).

3. Amber, borosilicate glass vials with PTFE-lined screw caps (10–25 mL, wide mouth) were used for storing drug and internal standard solutions at −15 to −30 °C.

4. Positive displacement pipets (Drummond Scientific Co. microdispensers), Hamilton syringes or Class A, glass, volumetric pipets were used for making drug and internal standard solutions.

5. SPEware Trace-B columns, 3 cc, 35 mg, mixed mode (C18/cation exchange) for solid-phase extraction.

6. HPLC columns (XTerra® MS C18, 3.5 µm, 2.1 × 100 mm) and guard columns (XTerra® MS C18, 5 µm, 2.1 × 10 mm) were purchased from Waters Co. (Milford, MA).

7. Pre-column filters (0.5 µm) were purchased from MAC-MOD Analytical, Inc. (Chadds Ford, PA).

**Table 1**
**Drugs present in the positive control (×10 dilution), their nominal concentrations, and typical Rt, RRt, and peak areas analyzed with the method as described**

| Drug | Concentration (μg/L) (ng/mL) | Typical Rt (min) and RRt | Typical (RRt) | Typical peak area |
|---|---|---|---|---|
| IS 1 (hydromorphone-D6) | 100 | 3.32 | 1.00 | 6.5E+04 |
| Acetaminophen | 400 | 3.57 | RRt1 = 1.10 | 2.7E+04 |
| Codeine | 160 | 4.20 | RRt2 = 0.35 | 8.0E+04 |
| Amphetamine | 160 | 4.13 | RRt2 = 0.36 | 1.2E+05 |
| Quinine | 160 | 5.24 | RRt2 = 0.44 | 5.9E+04 |
| Benzoylecgonine | 160 | 5.26 | RRt2 = 0.46 | 1.0E+05 |
| Phencyclidine | 80 | 6.56 | RRt2 = 0.56 | 7.1E+05 |
| EDDP (methadone metabolite) | 160 | 7.40 | RRt2 = 0.63 | 1.1E+06 |
| Imipramine | 80 | 7.71 | RRt2 = 0.66 | 1.0E+06 |
| Propoxyphene | 320 | 7.99 | RRt2 = 0.67 | 3.9E+05 |
| Methadone | 160 | 8.01 | RRt2 = 0.68 | 2.4E+06 |
| Oxazepam | 160 | 8.46 | RRt2 = 0.72 | 1.1E+05 |
| IS 2 (diazepam-D6) | 100 | 11.80 | 1.00 | 1.2E+05 |

**2.5 Equipment**

1. A Shimadzu Prominence HPLC system including (a) inline degasser, (b) two binary pumps with four solvent reservoirs and low-pressure solvent switching valves for each pump, (c) mixing T for gradient analysis, (d) temperature-controlled autosampler, (e) thermal column compartment with column switching valve.

2. An AB Sciex 3200 QTrap® hybrid triple-quadrupole linear ion trap mass spectrometer, including a 6-port divert valve.

3. Cliquid® software from Ab Sciex was used for batch instrument control and automated library search and reporting. Analyst® software from AB Sciex was used for instrument control during maintenance and manual library searching.

4. A 48 position positive pressure manifold from SPEware Co. was used for solid phase extraction (tubes).

5. A Turbovap LV (50 tubes) from Biotage was used to evaporate SPE eluates.

## 3   Methods

### 3.1   Stepwise Procedure

1. Sample Preparation
   (a) Centrifuge all urine samples at $3000 \times g$ for 10 min.
   (b) Label $13 \times 100$ mm glass tubes for control and patient samples.
   (c) Add 100 µL of the Mixed Working Internal Standard to all tubes using a positive displacement repeater pipettor.
   (d) Add 0.5 mL of the negative control, positive control, and patient samples to the appropriate tubes. Run one negative control and one positive control at the beginning of the run. Re-inject the negative control at the end of all runs. Re-inject the positive control at the end of the run if there are more than ten samples in the batch.
   (e) Add 1 mL of 0.1 M phosphate buffer (pH = 6.0) to all tubes. Vortex mix.

2. SPE Column Conditioning and Sample Extraction
   Note: Adjust pressure to allow a flow of approximately 20 drops/min from the column during all steps except drying.
   (a) Condition columns with 2 mL of methanol.
   (b) Condition columns with 1 mL of deionized water.
   (c) Transfer the samples into the SPEware Trace-B columns and elute samples through the columns.

3. Column Wash
   (a) Wash each column with 2 mL of deionized water.
   (b) Dry columns at maximum flow (50 psi for 10 min).

4. Elution, Drying, and Reconstitution
   (a) Elute drugs from the columns by applying 2.0 mL elution solvent (ethyl acetate:isopropanol: $NH_4OH$ at 73:25:2 vol:vol) to each column with flow into $13 \times 100$ mm glass tubes (make elution solvent fresh daily—*see* **Note 3**).
   (b) Evaporate to dryness for 20 min at 37 °C on the Turbovap.
   (c) Reconstitute with 0.5 mL Reconstitution Solvent. Vortex-mix thoroughly. Transfer the contents to autosampler vials using glass transfer pipets, and then cap the vials.

### 3.2   Instrument Operating Conditions

1. The LC and MS operating conditions are given in Tables 2, 3, and 4 (*see* **Note 4**).

2. A standardized equilibration procedure for the LC is necessary to insure reproducible Rts and peak areas for early eluting compounds (*see* **Note 5**).

3. The first two injections of any run are made with a short gradient (8 min) of 3–100 % B with flow directed to waste (via

**Table 2**
**LC operating conditions**

| Component | Parameter | Details |
|---|---|---|
| Binary pump | Gradient program | Time  Flow rate (μL/min)  %B<br>0.0   350   3<br>1.0   350   3<br>6.0   350   40<br>10.0   350   40<br>11.5   350   60<br>12.3   350   100<br>14.5   350   100<br>15.5   350   3<br>19.0   350   3 |
| Autosampler | Temperature | 10 °C |
| Autosampler | Volume injected | 10 μL (use 1 μL to identify very large peaks [>1e+6 area] or 30 μL for small peaks [<5e+3 area]) |
| Column oven | Temperature | 40 °C |
| Divert valve | Program | Time 0 = waste<br>Time 0.5 min = source<br>Time 15 min = waste |

divert valve) instead of to the source. The Rt of early eluting compounds had unacceptable variance unless these "LC primer" injections were used (*see* **Note 6**).

4. Procedures are necessary to avoid "collapse" of the XTerra® MS C18 stationary phase from exposure to low %B (e.g., 97:3 Mobile Phase A:Mobile Phase B) under certain conditions (*see* **Note 7**).

5. The guard column is changed at 500 injections or earlier if Rts, peak areas, or peak shapes are degraded.

6. A column/system wash procedure that removes the acidic buffer and leaves the instrument in a high percent acetonitrile milieu is used at the end of each run. This protocol has been found to extend the lifetime of the column to several thousand injections (*see* **Note 8**).

*3.3  Data Analysis*

1. The data is analyzed with Cliquid and Analyst software from AB Sciex using two printed reports: (1) a modified CES (collision energy spread) Best Candidate Detailed report and (2) a modified CES Confirmation Summary report. A portion (extracted ion chromatogram and library search summary list) from the CES Confirmation Summary report for a patient sample analyzed with this method is shown in Fig. 1.

**Table 3**
**MS operating conditions**

| Parameter | Setting | Details |
|---|---|---|
| Synchronization mode and duration | LC Sync 19 min | n/a |
| MRM experiment | Q1 mass, Q3 mass, CE, and CEP for 125 transitions | *See* Table 4 |
| MRM experiment | Ionization mode and dwell time | Positive ESI and 5 ms |
| MRM experiment | Advanced MS | Unit resolution for Q1 and Q3 |
| MRM experiment | Source/gas | Cur 20, CAD High, IS 4000, Temp 500, GS1 40, GS2 55, Ihe ON |
| IDA criteria | First level criteria | Select 1–3 most intense peaks which exceed 1000 cps, exclude former target ions after 3 occurrences for 15 s<br>Dynamic background subtraction: yes |
| 3 EPI experiments | MS | Scan speed 4000 Da/s, mass range 50–700 |
| 3 EPI experiments | Advanced MS | Q0 Trap OFF, LIT fill time of 50 ms, Dynamic Fill Time: On, TIC Target: $10.00 \times 1e + 7$, Minimum Fill time 2 ms, Maximum Fill Time 50 ms, Q1 unit resolution |
| 3 EPI experiments | Compound | DP 40, EP 10, CEP 9.08, CE 35, CE spread 15 |

2. The batch is considered acceptable if:
   (note peak areas used to evaluate QC and report patient samples are from the SRM survey scan, not from the EPI dependent scan):

   (a) For the negative control: No drugs are identified other than the internal standard, caffeine, theophylline (caffeine metabolite), and theobromine (caffeine metabolite). Both internal standards are present with peak areas above their thresholds, Rts within range, purities from library search ≥60 (*see* **Note 9**), and all spectra passing visual review. The baseline in the composite XIC is <500 cps (composite XIC is overlayed XICs of all 125 SRMs). Any unidentified peaks (no match or match <70) with areas >1e + 6 should be searched manually in Analyst Explore. A very large (>1e + 6) contaminant peak can be missed by the automated library search program, producing a false negative.

   (b) For the positive control: Both internal standards are present with peak areas above their thresholds, Rts within range, purities from library search ≥60, and all spectra

**Table 4**
**MRM parameters and verified detection limits for 125 compounds in the method**

| Q1 | Q3 | ID | DP | EP | CEP | CE | CXP | Verified at (ng/mL) |
|---|---|---|---|---|---|---|---|---|
| 333.1 | 211.2 | 2-Hydroxyethylflurazepam RRt2 = 0.75 | 40 | 10 | 24.82 | 50 | 5 | 100 |
| 328.1 | 165.1 | 6-MAM RRt1 = 1.36 | 40 | 10 | 24.68 | 35 | 5 | 100 |
| 286.1 | 121.1 | 7-Aminoclonazepam RRt1 = 1.71 | 56 | 4.5 | 16 | 39 | 4 | 100 |
| 284.1 | 135.1 | 7-Aminoflunitrazepam RRt2 = 0.53 | 56 | 4.5 | 16 | 35 | 4 | 100 |
| 427.2 | 207.1 | 9-Hydroxyrisperidone | 40 | 10 | 27.45 | 35 | 4 | n/d |
| 152.0 | 110.0 | Acetaminophen RRt1 = 1.10 | 40 | 10 | 19.75 | 20 | 5 | 5000 |
| 325.1 | 297.1 | alpha-Hydroxyalprazolam RRt2 = 0.70 | 66 | 6.5 | 19.2 | 35 | 4 | 100 |
| 359.0 | 330.8 | alpha-Hydroxytriazolam RRt2 = 0.70 | 61 | 9.5 | 20.2 | 35 | 4 | 100 |
| 309.1 | 281.1 | Alprazolam RRt2 = 0.75 | 56 | 6.5 | 18.8 | 35 | 5 | 100 |
| 646.0 | 100.0 | Amiodarone RRt2 = 1.12 | 40 | 10 | 33.58 | 50 | 5 | 500 |
| 278.2 | 117.2 | Amitriptyline RRt2 = 0.67 | 46 | 5.5 | 14 | 35 | 4 | 100 |
| 409.1 | 238.2 | Amlodipine | 15 | 3 | 18 | 15 | 3.5 | n/d |
| 314.1 | 193.1 | Amoxapine RRt2 = 0.62 | 40 | 10 | 18.36 | 50 | 5 | 100 |
| 136.1 | 91.1 | Amphetamine RRt1 = 1.32 | 21 | 7.5 | 8 | 25 | 4 | NR |
| 267.2 | 145.1 | Atenolol RRt1 = 1.11 | 40 | 10 | 22.97 | 35 | 5 | 100 |
| 290.2 | 124.1 | Atropine | 40 | 10 | 23.62 | 50 | 5 | n/d |
| 214.1 | 151.1 | Baclofen | 21 | 4 | 13 | 23 | 4 | n/d |
| 290.1 | 168.1 | Benzoylecgonine RRt1 = 1.62 | 40 | 10 | 23.61 | 35 | 5 | 100 |
| 308.2 | 167.1 | Benztropine | 40 | 10 | 24.12 | 50 | 5 | n/d |
| 319.1 | 274.1 | Brompheniramine RRt2 = 0.57 | 40 | 10 | 24.42 | 35 | 5 | 500 |
| 468.3 | 55.1 | Buprenorphine RRt2 = 0.61 | 86 | 10 | 24 | 85 | 4 | n/d |
| 240.0 | 183.9 | Bupropion RRt2 = 0.52 | 31 | 4 | 16.8 | 17 | 4 | 500 |
| 386.2 | 122.1 | Buspirone RRt2 = 0.54 | 40 | 10 | 26.3 | 35 | 5 | 100 |
| 195.1 | 122.9 | Caffeine | 40 | 10 | 20.95 | 50 | 5 | 5000 |
| 251.1 | 180.2 | Carbamazepine metabolite RRt2 = 0.59 | 40 | 10 | 16.38 | 50 | 5 | 500 |
| 237.1 | 193.1 | Carbamazepine RRt2 = 0.67 | 40 | 10 | 22.13 | 50 | 5 | 100 |
| 261.2 | 62.0 | Carisoprodol | 40 | 10 | 16.39 | 50 | 5 | 5000 |
| 389.2 | 201.1 | Cetirizine RRt2 = 0.72 | 40 | 10 | 26.39 | 20 | 5 | 500 |
| 300.1 | 227.1 | Chlordiazepoxide RRt2 = 0.56 | 40 | 10 | 23.92 | 20 | 5 | 500 |
| 275.1 | 230.1 | Chlorpheniramine RRt2 = 0.53 | 40 | 10 | 23.19 | 20 | 5 | 500 |

(continued)

**Table 4**
**(continued)**

| Q1 | Q3 | ID | DP | EP | CEP | CE | CXP | Verified at (ng/mL) |
|---|---|---|---|---|---|---|---|---|
| 319.1 | 86.1 | Chlorpromazine RRt2 = 0.70 | 40 | 10 | 24.42 | 20 | 5 | 500 |
| 253.1 | 95.1 | Cimetidine RRt1 = 1.09 | 40 | 10 | 16.45 | 35 | 5 | 500 |
| 325.2 | 109.0 | Citalopram RRt2 = 0.60 | 51 | 6.5 | 19.2 | 41 | 4 | 100 |
| 315.2 | 86.1 | Clomipramine RRt2 = 0.73 | 40 | 10 | 24.32 | 35 | 5 | 100 |
| 316.1 | 270.1 | Clonazepam RRt2 = 0.75 | 61 | 4.5 | 19 | 35 | 4 | 100 |
| 230.0 | 213.0 | Clonidine RRt1 = 1.21 | 40 | 10 | 21.93 | 35 | 5 | 100 |
| 327.1 | 270.1 | Clozapine RRt2 = 0.57 | 56 | 5 | 16 | 35 | 5 | 100 |
| 304.1 | 182.1 | Cocaine RRt1 = 1.80 | 41 | 6 | 16 | 21 | 4 | 100 |
| 300.2 | 152.2 | Codeine RRt1 = 1.25 | 56 | 7.5 | 14 | 85 | 5 | 100 |
| 177.1 | 80.0 | Cotinine RRt1 = 0.48 | 40 | 10 | 14.06 | 50 | 5 | 500 |
| 276.2 | 215.1 | Cyclobenzaprine RRt2 = 0.65 | 40 | 10 | 17.17 | 50 | 5 | 100 |
| 290.1 | 154.1 | D5-diazepam | 40 | 10 | 17.31 | 40 | 5 | I.S. 2 |
| 292.2 | 185.2 | D6-hydromorphone | 70 | 10 | 17.34 | 41 | 3 | I.S. 1 |
| 267.2 | 72.1 | Desipramine RRt2 = 0.65 | 36 | 4.5 | 16 | 25 | 4 | 100 |
| 272.2 | 213.4 | Dextromethorphan RRt2 = 0.58 | 76 | 3.5 | 30 | 35 | 4 | 100 |
| 258.2 | 157.3 | Dextrorphan RRt1 = 1.60 | 66 | 3 | 14 | 47 | 4 | 100 |
| 285.1 | 193.1 | Diazepam RRt2 = 1.02 | 56 | 5.5 | 12 | 41 | 4 | 100 |
| 302.2 | 199.1 | Dihydrocodeine RRt1 = 1.19 | 40 | 10 | 23.95 | 35 | 5 | 100 |
| 415.2 | 178.1 | Diltiazem RRt2 = 0.62 | 40 | 10 | 27.12 | 35 | 5 | 100 |
| 256.2 | 167.0 | Diphenhydramine RRt2 = 0.59 | 21 | 4 | 17.2 | 17 | 4 | 100 |
| 280.2 | 107.1 | Doxepin RRt2 = 0.61 | 41 | 4.5 | 18 | 31 | 4 | 100 |
| 271.2 | 167.1 | Doxylamine RRt1 = 1.57 | 40 | 10 | 17.02 | 35 | 5 | 100 |
| 200.1 | 182.1 | Ecgoninemethylester RRt1 = 0.28 | 40 | 10 | 21.09 | 35 | 5 | 500 |
| 278.2 | 186.2 | EDDP RRt2 = 0.63 | 66 | 5.5 | 14 | 39 | 4 | 100 |
| 166.1 | 91.1 | Ephedrine/pseudoephedrine RRt1 = 1.17 | 40 | 10 | 20.14 | 50 | 5 | 500 |
| 337.2 | 188.2 | Fentanyl RRt2 = 0.60 | 40 | 10 | 24.93 | 20 | 5 | 100 |
| 415.1 | 398.1 | Flecainide RRt2 = 0.61 | 40 | 10 | 27.11 | 20 | 5 | 100 |
| 314.1 | 268.1 | Flunitrazepam RRt2 = 0.81 | 56 | 4.5 | 18.9 | 35 | 4 | 100 |
| 310.1 | 148.1 | Fluoxetine | 40 | 10 | 24.17 | 35 | 5 | n/d |
| 388.1 | 315.1 | Flurazepam RRt2 = 0.60 | 40 | 10 | 26.36 | 20 | 5 | 100 |

(continued)

**Table 4**
**(continued)**

| Q1 | Q3 | ID | DP | EP | CEP | CE | CXP | Verified at (ng/mL) |
|---|---|---|---|---|---|---|---|---|
| 319.2 | 200.2 | Fluvoxamine RRt2 = 0.66 | 40 | 10 | 24.43 | 20 | 5 | 500 |
| 172.1 | 67.1 | Gabapentin RRt1 = 1.20 | 40 | 10 | 20.31 | 50 | 5 | 5000 |
| 446.2 | 321.2 | Glipizide RRt2 = 0.81 | 40 | 10 | 27.98 | 35 | 5 | 1000 |
| 376.1 | 123.0 | Haloperidol RRt2 = 0.61 | 40 | 10 | 26.02 | 50 | 5 | 100 |
| 300.2 | 199.1 | Hydrocodone RRt1 = 1.39 | 40 | 10 | 17.93 | 50 | 5 | 100 |
| 286.1 | 185.1 | Hydromorphone RRt1 = 1.02 | 40 | 10 | 17.48 | 50 | 5 | 100 |
| 375.2 | 201.1 | Hydroxyzine RRt2 = 0.66 | 40 | 10 | 26 | 20 | 5 | 100 |
| 281.2 | 86.1 | Imipramine RRt2 = 0.66 | 36 | 3.5 | 16 | 23 | 4 | 100 |
| 238.1 | 125.0 | Ketamine RRt1 = 1.49 | 40 | 10 | 22.16 | 35 | 5 | 100 |
| 329.2 | 91.0 | Labetalol RRt2 = 0.53 | 40 | 10 | 18.84 | 50 | 5 | 100 |
| 256.0 | 108.8 | Lamotrigine RRt1 = 1.57 | 56 | 9 | 17.2 | 67 | 4 | 100 |
| 235.2 | 86.1 | Lidocaine RRt1 = 1.48 | 40 | 10 | 22.08 | 20 | 5 | 100 |
| 477.2 | 266.2 | Loperamide | 40 | 10 | 28.85 | 50 | 5 | n/d |
| 321.0 | 275.0 | Lorazepam RRt2 = 0.75 | 51 | 5.5 | 19.1 | 29 | 4 | 100 |
| 278.2 | 250.2 | Maprotiline RRt2 = 0.66 | 40 | 10 | 23.28 | 20 | 5 | 100 |
| 180.1 | 163.2 | MDA RRt1 = 1.36 | 36 | 10.5 | 10 | 13 | 4 | 500 |
| 208.1 | 77.1 | MDE RRt1 = 1.52 | 40 | 10 | 21.32 | 50 | 5 | 100 |
| 194.1 | 163.2 | MDMA RRt1 = 1.42 | 26 | 4.5 | 14 | 15 | 4 | 100 |
| 248.1 | 220.1 | Meperidine RRt1 = 1.84 | 51 | 5 | 16 | 29 | 8 | 100 |
| 219.1 | 55.0 | Meprobamate | 40 | 10 | 15.38 | 50 | 5 | 5000 |
| 212.1 | 91.1 | Mescaline | 40 | 10 | 21.43 | 50 | 5 | n/d |
| 150.1 | 91.1 | Methamphetamine RRt1 = 1.38 | 31 | 10 | 13 | 25 | 4 | NR |
| 130.1 | 68.1 | Metformin RRt1 = 0.28 | 40 | 10 | 19.13 | 50 | 5 | >50,000 |
| 310.2 | 265.2 | Methadone RRt2 = 0.68 | 31 | 4 | 14 | 19 | 4 | 100 |
| 234.1 | 84.1 | Methylphenidate RRt1 = 1.68 | 40 | 10 | 22.04 | 35 | 5 | 100 |
| 268.2 | 103.0 | Metoprolol RRt1 = 1.67 | 40 | 10 | 23 | 50 | 5 | 100 |
| 266.2 | 195.1 | Mirtazapine RRt1 = 1.69 | 40 | 10 | 22.94 | 35 | 5 | 100 |
| 286.1 | 201.1 | Morphine RRt1 = 0.58 | 40 | 10 | 23.5 | 35 | 5 | 500 |
| 462.2 | 286.2 | Morphine-3-b-d-glucuronide RRt1 = 0.33 | 40 | 10 | 28.43 | 50 | 5 | >50,000 |
| 163.1 | 132.1 | Nicotine | 40 | 10 | 20.06 | 20 | 5 | 500 |

(continued)

**Table 4**
**(continued)**

| Q1 | Q3 | ID | DP | EP | CEP | CE | CXP | Verified at (ng/mL) |
|---|---|---|---|---|---|---|---|---|
| 347.1 | 254.1 | Nifedipine RRt2 = 0.93 | 40 | 10 | 25.21 | 50 | 5 | 1000 |
| 414.2 | 83.2 | Norbuprenorphine RRt2 = 0.54 | 86 | 4.5 | 20 | 67 | 4 | 5000 |
| 271.1 | 140.1 | Nordiazepam RRt2 = 0.81 | 56 | 4.5 | 17.7 | 37 | 4 | 100 |
| 266.1 | 107.2 | Nordoxepin RRt2 = 0.62 | 41 | 4 | 16 | 29 | 4 | n/d |
| 233.2 | 84.0 | Norfentanyl RRt1 = 1.55 | 41 | 5 | 14 | 25 | 4 | 100 |
| 234.1 | 160.1 | Normeperidine RRt2 = 0.52 | 46 | 5.5 | 12 | 19 | 4 | 100 |
| 302.1 | 187.1 | Noroxycodone RRt1 = 1.30 | 41 | 8.5 | 17.74 | 33 | 4 | 100 |
| 308.2 | 100.1 | Norpropoxyphene RRt2 = 0.66 | 26 | 4.5 | 18 | 19 | 4 | 1000 |
| 264.2 | 233.1 | Nortriptyline RRt2 = 0.66 | 41 | 4.5 | 16 | 19 | 4 | 100 |
| 313.1 | 256.1 | Olanzapine RRt1 = 1.33 | 40 | 10 | 24.26 | 20 | 5 | 100 |
| 287.1 | 241.1 | Oxazepam RRt2 = 0.72 | 46 | 4.5 | 18.2 | 31 | 4 | 100 |
| 253.1 | 236.0 | Oxcarbazepine RRt2 = 0.61 | 40 | 10 | 22.58 | 20 | 5 | n/d |
| 316.1 | 298.1 | Oxycodone RRt1 = 1.33 | 41 | 4 | 14 | 25 | 6 | 500 |
| 302.1 | 198.1 | Oxymorphone RRt1 = 0.77 | 40 | 10 | 17.99 | 50 | 5 | 500 |
| 330.1 | 192.1 | Paroxetine RRt2 = 0.64 | 40 | 10 | 24.73 | 50 | 5 | 500 |
| 286.2 | 218.3 | Pentazocine RRt2 = 0.55 | 70 | 10 | 17.19 | 25 | 3 | 100 |
| 244.1 | 91.0 | Phencyclidine RRt2 = 0.56 | 31 | 5 | 15.86 | 37 | 4 | 100 |
| 150.1 | 91.2 | Phentermine RRt1 = 1.48 | 70 | 10 | 12.88 | 28 | 3 | 5000 |
| 152.1 | 134.0 | Phenylpropanolamine RRt1 = 0.92 | 40 | 10 | 13.28 | 50 | 5 | 1.000 |
| 285.1 | 86.1 | Promethazine RRt2 = 0.62 | 40 | 10 | 17.45 | 35 | 5 | 100 |
| 340.2 | 58.1 | Propoxyphene | 31 | 4.5 | 22 | 31 | 4 | 300 |
| 260.2 | 183.1 | Propranolol RRt2 = 0.56 | 40 | 10 | 22.78 | 20 | 5 | 100 |
| 384.2 | 253.1 | Quetiapine RRt2 = 0.59 | 51 | 7.5 | 20.8 | 29 | 4 | 100 |
| 325.2 | 307.2 | Quinidine/quinine RRt1 = 1.56 | 40 | 10 | 24.6 | 35 | 5 | 100 |
| 315.2 | 176.1 | Ranitidine RRt1 = 1.12 | 40 | 10 | 24.32 | 35 | 5 | 100 |
| 411.2 | 191.1 | Risperidone RRt2 = 0.53 | 40 | 10 | 27 | 35 | 5 | 100 |
| 306.1 | 159.0 | Sertraline RRt2 = 0.71 | 40 | 10 | 24.06 | 35 | 5 | 500 |
| 301.1 | 255.1 | Temazepam RRt2 = 0.83 | 41 | 5 | 18.6 | 29 | 4 | 100 |
| 181.1 | 124.0 | Theophylline RRt1 = 1.25 | 40 | 10 | 20.56 | 20 | 5 | 2000 |
| 264.2 | 58.0 | Tramadol | 26 | 4 | 17.4 | 35 | 4 | 100 |

(continued)

**Table 4**
**(continued)**

| Q1 | Q3 | ID | DP | EP | CEP | CE | CXP | Verified at (ng/mL) |
|---|---|---|---|---|---|---|---|---|
| 372.2 | 176.1 | Trazodone RRt2 = 0.54 | 40 | 10 | 25.91 | 35 | 5 | 100 |
| 343.0 | 239.0 | Triazolam RRt2 = 0.78 | 40 | 10 | 25.09 | 50 | 5 | 100 |
| 278.2 | 260.2 | Venlafaxine RRt2 = 0.53 | 40 | 10 | 23.3 | 20 | 5 | 100 |
| 455.3 | 165.2 | Verapamil RRt2 = 0.66 | 40 | 10 | 28.24 | 35 | 5 | 100 |
| 308.2 | 235.2 | Zolpidem RRt2 = 0.52 | 66 | 5 | 16 | 45 | 4 | 100 |

*n/d* not done

passing visual review. All expected compounds are present with peak areas above their thresholds, relative retention times (RRts) within range (*see* **Note 10**), Purities for library searches ≥70, and all spectra passing visual review (*see* **Note 11**). Only expected compounds (11 spiked drugs and 2 internal standards) are found.

3. Drugs/metabolites in patient samples are reportable if:

(a) Both internal standards are present in the sample with peak areas above their thresholds, Rts within range, purities from library search ≥60, and all spectra passing visual review (*see* **Note 12**).

(b) The drug/metabolite(s) have a peak area(s) >5e + 3, relative retention times (RRts) within range, purities from library search ≥70, and all spectra pass visual review. An example of the Best Candidate Detailed Report for one EPI scan is shown in Fig. 2.

(c) Visual review of the match between each unknown spectra and library spectra using in-house-developed criteria is necessary to avoid false positives [5, 9].

4. Manual library search (using Analyst Explore function) is necessary in some cases to avoid false negatives (*see* **Note 13**). An example of manual library search results from Analyst Explore for chlordiazepoxide is shown in Fig. 3.

5. Carryover was evaluated during validation. Infrequently peaks with area >4e + 6 caused carryover to the next specimen. Carryover appeared to be drug dependent.

6. Results for limit of detection studies are shown in Table 4.

## Extracted Ion Chromatogram

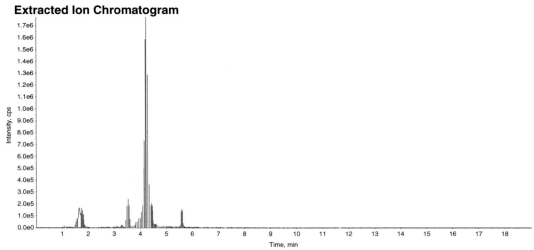

## Summary

| RT (min) | Exp RT (min) | RRt | Compound Name | Peak Area | Purity (%) | Fit (%) | Rev Fit (%) |
|---|---|---|---|---|---|---|---|
| 1.07 | 1.07 | 0.33 | Nicotine RRt1=0.34 | 1.45e+005 | 88.0 | 94.1 | 93.5 |
| 1.66 | 1.57 | 0.50 | Cotinine RRt1=0.48 | 2.33e+006 | 96.6 | 100.0 | 96.6 |
| 2.78 | 2.35 | 0.84 | Oxymorphone RRt1=0.77 | 1.38e+005 | 89.4 | 95.7 | 93.4 |
| 3.52 | 2.95 | 1.06 | Acetaminophen RRt1=1.10 | 2.48e+004 | 93.4 | 100.0 | 93.4 |
| 3.29 | 3.22 | 1.00 | D6-Hydromorphone RRt2=0.29 | 1.62e+005 | 73.1 | 75.4 | 97.0 |
| 3.68 | 3.24 | 1.11 | Hydromorphone RRt1=1.02 | 1.23e+005 | 42.9 | 94.9 | 45.2 |
| 3.54 | 3.57 | 1.07 | Acetaminophen RRt1=1.10 | 1.90e+006 | 96.9 | 98.9 | 98.0 |
| 3.43 | 3.71 | 1.04 | Ephedrine/Pseudoephedrine RRt1=1.17 | 4.62e+004 | 44.6 | 62.6 | 71.3 |
| 3.91 | 3.99 | 0.33 | Theophylline RRt2=0.34 (RRt1=1.25) | 5.18e+005 | 81.6 | 96.0 | 85.0 |
| 4.15 | 4.07 | 0.35 | Noroxycodone RRt2= 0.37 (RRt1=1.30) | 9.64e+005 | 81.4 | 92.6 | 87.9 |
| 4.66 | 4.13 | 0.40 | Amphetamine RRt2=0.38 (RRt1=1.28) | 4.14e+004 | 80.7 | 80.7 | 100.0 |
| 4.44 | 4.20 | 0.38 | Hydrocodone RRt2=0.38 (RRt1=1.39) | 5.32e+005 | 91.5 | 97.0 | 94.4 |
| 4.22 | 4.23 | 0.36 | Oxycodone RRt2=0.37 (RRt1=1.33) | 1.78e+007 | 79.1 | 98.3 | 80.5 |
| 4.16 | 4.37 | 0.35 | Noroxycodone RRt2= 0.37 (RRt1=1.30) | 4.36e+006 | 81.4 | 92.6 | 87.9 |
| 4.43 | 4.50 | 0.38 | Hydrocodone RRt2=0.38 (RRt1=1.39) | 1.37e+006 | 91.5 | 97.0 | 94.4 |
| 4.52 | 4.64 | 0.38 | Caffeine RRt2=0.40 (RRt1=1.45) | 7.37e+004 | 94.4 | 100.0 | 94.4 |
| 5.60 | 5.62 | 0.48 | 7-Aminoclonazepam RRt2=0.48 (RRt1=1.77) | 9.31e+005 | 90.7 | 92.7 | 97.9 |
| 8.67 | 8.77 | | Library search indicates no matching spectra found. | | | | |
| 4.06 | 9.65 | | Library search indicates no matching spectra found. | | | | |
| 11.70 | 11.80 | 1.00 | D5-Diazepam RRt1=3.66 | 3.58e+005 | 77.7 | 85.3 | 91.1 |

**Fig. 1** The Composite Extracted Ion Chromatogram (EIC or XIC) and the Cliquid® CES Confirmation Summary Report for a patient specimen analyzed with this method are shown. All EPI spectra that were searched are matched to the highest purity hit from the library and listed in Rt order

## 4 Notes

1. Minimize exposure of all organic solvents used in the method (in standards, internal standards, mobile phases, pump washes, etc.) to any form of plastic (containers, vials, pipets, transfer pipets, parafilm, pipet tips, etc.). Phthalate contamination from plastics is

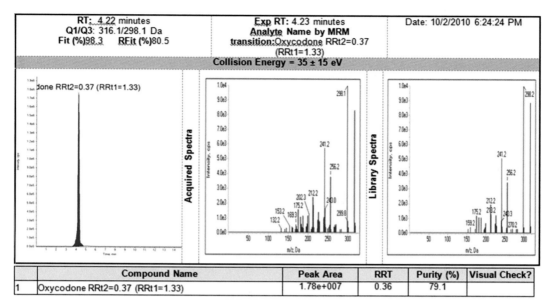

| | Compound Name | Peak Area | RRT | Purity (%) | Visual Check? |
|---|---|---|---|---|---|
| 1 | Oxycodone RRt2=0.37 (RRt1=1.33) | 1.78e+007 | 0.36 | 79.1 | |

**Fig. 2** An example showing oxycodone identified in the Cliquid® CES Best Candidate Detailed Report for the patient sample from Fig. 1. The *left pane* is the SRM peak found in the survey scan of the patient sample. The *center pane* is the EPI spectrum from a dependent scan of the patient sample. The *right pane* is the reference EPI spectrum from the 125 compound library. The highest purity hit from library search is listed below with the observed SRM peak area, observed RRt and purity. The naming convention for compounds added to the in-house library included expected Rt and RRt that had been established in-house with reference compounds. This allowed for easy comparison with the observed Rt and RRt of candidate compounds in patient specimens. RRt1 indicates that Internal Standard 1, hydromorphone-d6, was used to calculate the relative retention time (compounds in the first 4 min of the run) and RRt2 indicates that Internal Standard 2, diazepam-d5, was used the calculate the relative retention time (compounds with Rts later than 4 min)

common and can be significant. Use polytetrafluoroethylene (PTFE) lined caps for containers and PTFE lined septa for ALS vials.

2. Two internal standards, the early eluting hydromorphone-D6 and the late-eluting diazepam-D6, are used because the Rts of early eluting drugs are more variable than the Rts of mid-run and late-eluting compounds. Use of the early-eluting hydromorphone-D6 as an internal standard more reliably gave RRts within the expected range for morphine, hydromorphone, and other drugs eluting in the first 4 min.

3. SPE elution solvent is made fresh each day of use because of the high volatility of $NH_4OH$.

4. A single mid-range energy collision energy (CE) does not yield diagnostic spectra for all compounds. Frequently, lower molecular weight ($<250 \, m/z$) compounds tend to be over fragmented when CE is too high in to non-diagnostic, low-molecular-weight product ions. Higher molecular weight drugs ($>400 \, m/z$) tend to be under fragmented when CE is

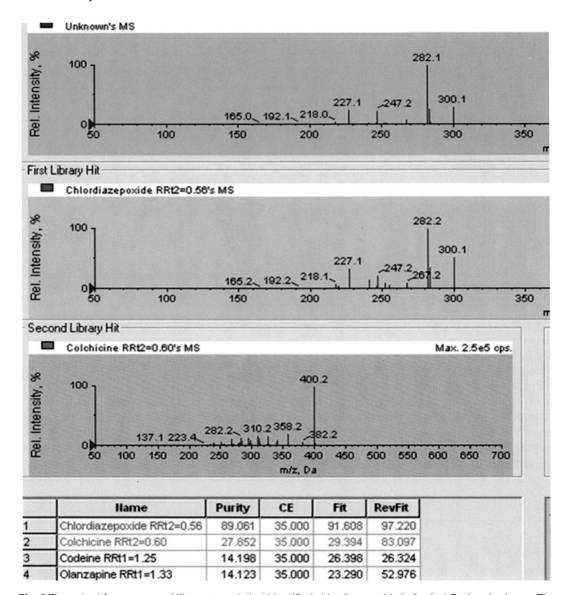

**Fig. 3** The output from a manual library search that identified chlordiazepoxide in Analyst Explore is shown. The *top pane* is the unknown EPI spectrum from the dependent scan of the patient sample. The *next pane down* is the reference EPI spectrum with the highest purity hit from the manual library search (chlordiazepoxide with Purity 89.1). The *next pane* is the second highest purity hit (colchicine with purity 27.8). The top four hits from the library search, ranked by purity, and showing the CE at which the library spectra were acquired, are listed in the *bottom pane*

too low, with few or no product ions. The QTrap® Collision Energy Spread (CES) function is used in this method to apply three user defined collision energies (20, 35, and 50), with product ions moved out of the collision cell and held in the LIT after each CE is applied, to better fragment the widest array of compounds into useful product ion spectra.

5. The method uses a weak buffer (2 mmol/L) and starts at a very low % of organic solvent (3 % mobile phase B or 2.7 % acetonitrile). These two factors cause the Rts of early eluting drugs to be very sensitive to incomplete equilibration of the LC system from the column wash conditions (100 % acetonitrile). Incomplete equilibration causes Rt and RRt shifts. Equilibration before the run was 2 min at 50 % mobile phase B and 13 min at 10 % mobile phase B.

6. Two "LC primer" injections are performed after LC equilibration as follows: 0.5–8.0 min gradient from 3 to 100 % B at 350 μL/min; 100 to 3 % B from 8.0 to 8.5 min at 350 μL/min; 3 min re-equilibration at 3 % B at 350 μL/min.

7. C18 stationary phases may "collapse" or "dewet" when exposed to a low percent organic component in the mobile phase (e.g., ≤10 %). We observed Rt shifts and bizarre peak shapes under these conditions. This problem occurs if the column used for this method is left in <10 % mobile phase B without flow or when equilibrated at isocratic flow conditions of <10 % B. It does not occur with a continuously cycling gradient starting at 3 % B. The equilibration and "primer" injection procedures described in Notes 5 and 6 are successful at preventing this phenomenon.

8. The "column/system wash" is performed at the end of a run. The divert valve is switched to waste and the column is washed at 50:50 acetonitrile:water for 5 min then 100 % acetonitrile for 20 min.

9. The internal standards required a library search Purity of only ≥60, instead of the ≥70 required for unknowns. If the internal standard identification was wrong or missing or the purity was <60—the internal standards were searched manually using Analyst Explore (see Note 11).

   If both I.S. peak areas in a patient sample were below threshold the sample was re-extracted. If the I.S. 2 peak area was acceptable but the I.S. 1 peak area was very low or missing, ion suppression was assumed. If I.S. 1 was missing, expected Rt instead of RRt could be used for verifying drugs that use I.S. 1 for RRt (Rts earlier than 4.0 min) and a comment warning of possible false negatives was appended. The acceptable range for expected Rt was ±0.2 min.

10. Relative retention time (RRt) criteria: RRt is acceptable if within ±0.05 or within 5 % of the library standard RRt—whichever is larger.

11. In-house-developed EPI Visual Spectral Review Criteria [5]: The tolerance for ions in the unknown spectra to match by visual review to ions in the library spectra was ±0.2 $m/z$. For example a product ion of 227.3 $m/z$ matches to a library

spectra ion of 227.1. An ion of 227.4 does not match. The four most abundant ions in the library spectra should be present in the unknown spectra. If there are ions in the unknown that are NOT present in the library spectra—they should be less abundant than the two most abundant ions in the unknown spectra OR if there are only three ions in the library spectra—all three ions must also be present in the unknown spectra. If there are more than three ions in the unknown— the "extra" ions—those that are not present in the library spectra—should be less abundant than the three matched ions AND there should be other evidence that supports a positive result. Other evidence includes a positive result for that drug with a different method (GC-MS or immunoassay) OR a positive result for a metabolite/parent of the drug in question with this method OR exceptions noted for selected spectra in the library. For example—9-hydroxyrisperidone, cetirizine, lidocaine, and sertraline have sparse spectra and so were reportable with only three ions. Selected drugs/drug classes did not reliably have three ions present in the EPI spectrum and were reported instead from targeted confirmation methods, e.g., amphetamine and methamphetamine.

If the purities from library search for drugs in the positive QC were too low (<70)—the spectra were checked for missing/decreased abundance of ions <100 $m/z$ compared to the library spectra. On a few occasions, this was a sign that the linear ion trap (LIT) needed to be tuned and re-calibrated.

12. Visual review criteria for the Internal Standards were different than for unknowns: Internal standard 1 was accepted with only the precursor ion (292) present so long as it was the most abundant ion. Internal standard 2 was accepted if only two ions (290 precursor ion and one other) were present so long as 290 was the most abundant ion and the other ion matched to an ion in the library spectrum.

13. Peaks with areas $\geq 1e+6$ and purities <70 or with "*-no matching spectra found-*" reported on the Best Candidate Detailed Report were potentially false negatives. The most common cause of false negatives was from too many ions present in the LIT causing space charging and mass bias (e.g., $m/z$ 278 appears as $m/z$ 279). For such large peaks, a spectrum can usually be searched manually in Analyst Explore from the shoulder or tail of the peak (lower abundance of ions in the LIT) or from the second or third EPI experiment and matched with a purity $\geq 70$. The sample extract can also be re-injected with a lower volume (e.g., 1 μL instead of 10 μL).

False negatives can also occur because too little drug is present for a satisfactory spectrum to be obtained. When the SRM peak is <1e+4, purities may be <70 and spectra may have less than four ions matching to the library spectra, even for a true positive. If manual search does not find an acceptable spectrum—the sample can be re-injected with more volume (e.g., 30 µL instead of 10 µL).

## Acknowledgements

The author would like to thank Julia Drees, Ph.D., Scientific Director, and Judy Chang, M.Sc., Scientific Director, from Kaiser TPMG Regional Laboratories, Berkeley, CA, and Christopher Borton, Ph.D., from AB Sciex for their contributions to the development of this method, expert advice, and fruitful collaboration.

## References

1. Maurer HH (2006) Hyphenated mass spectrometric techniques-indispensable tools in clinical and forensic toxicology and in doping control. J Mass Spectrom 4:1399–1413, Review

2. Marquet P (2002) Is LC-MS suitable for a comprehensive screening of drugs and poisons in clinical toxicology? Ther Drug Monit 24:125–133, Review

3. Saint-Marcoux F, Lachâtre G, Marquet P (2003) Evaluation of an improved general unknown screening procedure using liquid chromatography-electrospray-mass spectrometry by comparison with gas chromatography and high-performance liquid-chromatography— diode array detection. J Am Soc Mass Spectrom 14:14–22

4. Sauvage FL, Saint-Marcoux F, Duretz B, Deporte D, Lachâtre G, Marquet P (2006) Screening of drugs and toxic compounds with liquid chromatography-linear ion trap tandem mass spectrometry. Clin Chem 52: 1735–1742

5. Lynch KL, Breaud AR, Vandenberghe H et al (2010) Performance evaluation of three liquid chromatography mass spectrometry methods for broad spectrum drug screening. Clin Chim Acta 411:1474–1481

6. Mueller CA, Weinmann W, Dresen S, Schreiber A, Gergov M (2005) Development of a multi-target screening analysis for 301 drugs using a QTrap liquid chromatography/tandem mass spectrometry system and automated library searching. Rapid Commun Mass Spectrom 19:1332–1338

7. Arhard K, Gottshcall A, Pitterl F et al (2015) Applying 'Sequential Windowed Acquisition of All Theoretical Fragment Ion Mass Spectra' (SWATH) for systematic toxicological analysis with liquid chromatography-high-resolution tandem mass spectrometry. Anal Bioanal Chem 407:405–414

8. Roemmelt AT, Steuer AE, Poetzsch M et al (2014) Liquid Chromatography, in combination with a quadrupole Time-of-Flight instrument (LC QTOF), with sequential windowed acquisition of all theoretical fragment ion mass spectra (SWATH) acquisition: systematic studies on its use for screenings in clinical and forensic toxicology and comparison with information-dependent acquisition (IDA). Anal Chem 86:11742–11749

9. Chase D et al (2007) Mass spectrometry in the clinical laboratory: general principles and guidance. Approved Guideline C50-A. Clinical and Laboratory Standards Institute, Wayne, PA

# Chapter 17

# High-Resolution Mass Spectrometry for Untargeted Drug Screening

## Alan H.B. Wu and Jennifer Colby

### Abstract

While gas chromatography-mass spectrometry (GC/MS) continues to be the forensic standard for toxicology, liquid chromatography coupled to tandem MS offers significant operational advantages for targeted confirmatory analysis. LC-high-resolution (HR)-MS has recently been available that offers advantages for untargeted analysis. HR-MS analyzers include the Orbitrap and time-of-flight MS. These instruments are capable of detecting 1 ppm mass resolution. Following soft ionization, this enables the assignment of exact molecular formula, limiting the number of candidate compounds. With this technique, presumptive identification of unknowns can be conducted without the need to match MS library spectra or comparison against known standards. For clinical toxicology, this can greatly expand on the number of drugs and metabolites that can be detected and reported on a presumptive basis. Definitive assignments of the compound's identity can be retrospectively determined with acquisition of the appropriate reference standard.

**Key words** High-resolution mass spectrometry, Orbitrap, Time-of-flight mass spectrometry, Untargeted mass screening, Exact molecular formula

## 1 Introduction

### 1.1 Mass Spectrometry

Mass spectrometry as a detector has played a prominent role in drug screening and confirmation analysis for many years beginning with gas chromatography-mass spectrometry (GC-MS) followed by liquid chromatography-mass spectrometry (LC-MS), GC-tandem mass spectrometry (GC-MS/MS), and LC-tandem mass spectrometry. Drugs are typically bombarded by electrons to form ions and to fragment the molecular ion into pieces of predictable size. The ions are sorted according to their mass to charge ratio ($m/z$) using a quadrupole filtering segment. As the name implies, the quadrupole consists of four cylindrical rods. Each pair is connected to a direct and radiofrequency voltage. Ions pass through the middle of these rods and form either unstable orbits which are lost to the vacuum, or stable orbits that reach the electron

Uttam Garg (ed.), *Clinical Applications of Mass Spectrometry in Drug Analysis: Methods and Protocols*, Methods in Molecular Biology, vol. 1383, DOI 10.1007/978-1-4939-3252-8_17, © Springer Science+Business Media New York 2016

multiplier detector. Due to the complexity of biological samples, most mass spectrometers are connected to a gas or liquid chromatograph, which isolates and purifies the sample, prior to its introduction to the mass spectrometer. Because the resolution of liquid chromatography is lower than for gas chromatography, the use of daughter ion spectra, derived from the parent ion as generated in a tandem mass spectrometer, is a means to further isolate the targeted compound of interest.

## 1.2 High-Resolution (HR) Mass Spectrometry (MS) Instrumentation

Time-of-flight (TOF) mass spectrometry (MS) instrumentation makes use of an alternative mass filtering device to the quadrupole. While GC-MS and LC-MS make use of electron impact and electrospray ionization, optimum use of TOF instrument requires "soft ionization" which minimizes parent ion fragmentation. Molecular ions are focused into a flight tube where time is measured from the point of ionization from the source, traveling through the flight tube, and when it reaches the detector. Lighter ions travel faster under a given electrical field than heavier ions. The major advantage of TOF MS analyzers is high mass resolution. The molecular weight of parent ions can be determined with a 2–5 part per million (ppm). From this, the exact molecular formula of a particular chromatographic peak can be determined in most cases. The traditional quadruple mass filtering analyzers are only able to differentiate nominal molecular weights, i.e., whole molecular mass units. There are many manufacturers of TOF MS instruments. The combination of a triple quadrupole with TOF MS/MS enables the simultaneous assessment of molecular fragmentation with precise molecular formula assessments.

There are other high-resolution mass spectrometers that have been developed to compete against time-of-flight instrumentation. The Orbitrap (manufactured by Thermo-Fisher, San Jose, CA) consists of two electrodes, the outer barrel-like electrode and an inner spindle-like electrode, which are used to create an electrostatic field. Ions from the chromatograph are focused tangentially into the electrical field where they undergo stable and unstable oscillations from one end of the inner electrode to the other. This configuration enables the production of high-resolution mass spectra that are equivalent to those of TOF MS instruments.

The resolution of HR MS instruments is defined as the absolute difference from the exact theoretical molecular mass to the measured mass divided by the theoretical mass. This ratio is multiplied by $10^6$ to produce a value in parts per million. Typically, TOF and Orbitrap instruments can produce a resolution of 2–5 ppm. For an $m/z$ of 100, a 2 ppm resolution produces a mass error of less than $m/z$ 0.0002. This is more than adequate to assign molecular formulae to candidate compounds. Once the formulae are known, the analyst can conduct a web-based internet searches using the formula as the keyword. Compounds will

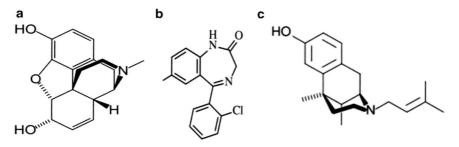

**Fig. 1** Differentiation of molecular formula by time-of-flight mass spectrometry. (**a**) Morphine: formula $C_{17}H_{19}NO_3$, molecular weight 285.3377. (**b**) 7-aminoclonazepam (major clonazepam metabolite): formula $C_{15}H_{12}ClNO_3$, molecular weight 285.7283. (**c**) Pentazocine: formula $C_{19}H_{27}NO$, molecular weight 285.2093

appear that will necessarily be isomers of each other (stereo or structural). The complexity of high resolution mass spectrometry increases with the mass-to-charge ratio. It is more difficult to assign the identity of a compound based on molecular weight as the size of the ion increases.

To illustrate the advantage of high-resolution mass spectrometry, Fig. 1 shows the comparison of three drugs and metabolites with the same nominal molecular weight, but have different molecular formula and thus different exact molecular weights. Using standard single or triple quadrupole MS, a single peak corresponding to the ion at $m/z$ 285 would appear with each compound. Using TOF-MS, the $m/z$ 285.34 ions would be separated from 285.73 from 285.21. Comparison of GC-MS, LC-MS/MS, and LC-HR/MS for small-molecule analysis is shown in Table 1.

*1.3 High-Resolution Mass Spectrometry for Clinical and Forensic Toxicology (Fig. 2)*

Currently, drug testing in clinical and forensic practice is a two-step process. The initial testing by use of immunoassays is followed by confirmatory analysis by mass spectrometry. Immunoassays are designed to detect one or more members of a class of drugs and/ or their metabolites. For example, the immunoassay for opiates detects codeine, morphine, hydrocodone, and hydromorphone. Confirmatory assays are necessary for many clinical and all forensic applications because some immunoassays suffer from false-positive results, i.e., unintended cross-reactivity towards related drugs and compounds, while other assays produce false-negative results, i.e., the inability of an immunoassay to detect all members of a class of drugs. For example, diphenhydramine produces a false-positive result on the phencyclidine assay [1] while most commercial immunoassays for benzodiazepines do not detect lorazepam or the lorazepam glucuronide (the major lorazepam metabolite) [2]. Targeted GC-MS or LC-MS is used to confirm results of immunoassay screening tests. If a member of the targeted drug class is present, a positive result is reported. If the positive result is due to an interfering drug or compound, a negative result is reported.

**Table 1**
**Comparison of tandem for targeted versus TO F/MS for nontargeted small-molecule analysis[a]**

| Parameter | GC-MS | LC-MS/MS | LC-HR/MS |
|---|---|---|---|
| Scan modes | Targeted (SIM)[b] | Targeted (SRM) | Full scan |
| Fragmentation requirement | Yes | Yes | No (parent ion only) |
| Mass resolution | Low (nominal mass) | Low (nominal mass) | High (0.001 amu) |
| Mass range | Low (<3000 amu) | Low (<3000 amu) | High (20,000 mu) |
| Identification criterion | SIM/ion ratios | SRM/ion ratios mass | Molecular formula |
| Databases | MS libraries | MS libraries | Accurate mass |
| Instrumentation costs | | | |
| Method development | Low ($50–100,000) | High ($200–400k) | High ($200–400k) |
| Method development | | | |
| For qualitative analysis | Extensive | Extensive | Abbreviated, nontargeted |

[a]*GC-MS* gas chromatography-mass spectrometry, *LC-MS/MS* liquid chromatography-tandem mass spectrometry, *LC-HR/MS* liquid chromatography-high-resolution mass spectrometry which includes time-of-flight and Orbitrap mass spectrometry. Used with permission from Wu et al. Clin Toxicol 2012;50:733–42
[b]*SIM* selected ion monitoring, *SRM* selected reaction monitoring

Mass spectrometry can also be used to quantitate the concentration of the drug in urine, serum, or other body fluids. Quantitative results are mandated by the Federal Workplace Drug Testing Programs for federal employees [3], although the concentration of the drug or metabolite is not reported. Quantitative testing is also important in postmortem (forensic) drug testing, as the levels can be helpful to determine if the drug present is incidental or a major cause of death.

Mass spectrometry can also be used as a screening assay for unknown drugs and intoxicants. This is important for drugs for which there are no immunoassays available. Drugs that fall into this category and have clinical relevance include designer amines and cannabinoids, novel antipsychotic drugs, oral hypoglycemic medications, and antiarrhythmias. A computerized search algorithm is used to match unknown peaks from a mass fragmentation pattern to a mass spectra library. Comprehensive libraries are available using electron impact ionization for GC-MS. These libraries are transportable; that is, there is consistency in the mass spectrum produced from different instruments. However, these libraries may not contain spectra for many of the known metabolites of targeted drugs. Derivatization prior to GC-MS analysis is necessary in order to produce a compound that can be injected into the gas chromatogram. LC-MS and LC-MS/MS have been used for general unknown screening. Unlike GC-MS, samples injected for LC-MS does not require prior derivatization of the sample, as the sample

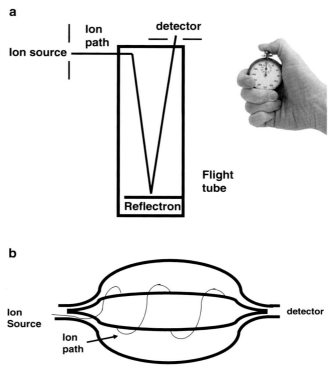

**Fig. 2** Simplified diagram of filtering strategies for high-resolution mass spectrometers. (**a**) Time-of-flight MS. Ions are focused onto a long flight tube. The time needed from ionization to detector is measured. Heavier ions take longer time than lighter ions. (**b**) Orbitrap MS. Ions follow a stable orbit around the trap and are selectively ejected from the orbits

does not need to be volatile. However, LC-MS assays for toxicology applications produce more variability in the mass spectrum between different instrument manufacturers and separate unknown libraries must be created.

LC time-of-flight mass spectrometers and other high resolution MS instruments have been examined as an alternative to tandem MS for unknown toxicology screening [4–8]. High-resolution MS enables determination of the exact molecular formula that can be useful for presumptive assignment of unknowns in a general toxicology screen, even if the toxicology laboratory has no prior experience with that drug. The laboratory may not have a standard, know the chromatographic retention time of the compound, or have the full-scan mass spectrum in their library. Reporting of presumptive results can be useful in clinical toxicology situations where turnaround time is essential in making management or triaging decision of patients who are acutely poisoned or overdosed. The clinical team, however, must be knowledgeable as to the limitation of untargeted screening procedures.

Fortunately, the concept of reporting "presumptive findings," i.e., the exact assignment of the intoxicant is not made, has been used for many years in conjunction with other non-definitive toxicology procedures such as thin-layer chromatography (e.g., Toxi® Lab, Analytical Systems, Laguna Hill, CA) [9] and liquid chromatography rapid UV scanning LC detection (e.g., Remedi® BioRad Laboratories, Benicia, CA) [10]. Both of these techniques have largely been abandoned in clinical toxicology laboratories today.

High-resolution mass spectrometry can be useful for forensic applications as a screening tool; however additional measures must be taken to confirm presumptive findings found. Once a drug or intoxicant is known, the laboratory must acquire the drug standards and validate the results. Alternately, the sample could be sent to a reference laboratory that has validated procedures for the drug in question. A presumptive result, even if produced by mass spectrometry, is not defendable in court. Figure 3 shows a possible flow of samples using high resolution mass spectrometry and the different approach taken for clinical versus forensic testing. Clinical testing favors rapid turnaround time while forensic analysis requires production of defensible results.

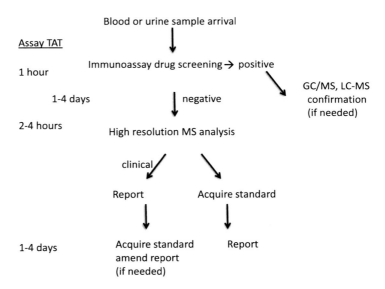

**Fig. 3** Flow of analytical assays for clinical and forensic toxicology analysis using mass spectrometry. Immunoassay results or turnaround time (TAT) can be obtained within 1 h after sample receipt. Positive immunoassay results can be confirmed by GC-MS or LC-MS/MS. Drugs not tested by immunoassay can be examined by high-resolution mass spectrometry. For clinical purposes, a presumptive results can be obtained within a few hours and reported immediately. For forensic purposes, definitive testing is needed to confirm before results can be reported should there be any legal proceedings

## 1.4 Applications in Drug Testing using High-Resolution Mass Spectrometry

There are many different toxicology questions that can be addressed with high-resolution MS analysis. Perhaps the strongest application is in the identification of novel designer drugs. These are analogs to exiting drugs whereby one or more functional group has been altered in such a way that the pharmacologic activity is similar to the parent drug. The objective of is to produce a drug that is similar to one that is either illegal or whose distribution is controlled by a country's drug enforcement agencies. Although designer drugs have been synthesized for illicit purposes for many decades, there is a resurgence of the synthesis of new compounds, particularly for amines such as "bath salts" [11], marijuana [12], and fentanyl [13]. Most of these designer amines are not detected by the existing commercial immunoassays [14], unless they are specifically targeted towards those compounds. The rate by which these compounds are produced by clandestine laboratories exceeds the ability of the toxicology laboratory to produce targeted mass spectrometric analysis, which typically takes several months to complete. At one time, it also greatly exceeds the rate by which the US Drug Enforcement Agency could schedule these drugs as "controlled substances." However, recent legislation does not require DEA to list specific drugs, in favor of classes of agents. For example, the term "cannabimimetic agents" means any substance that is a cannabinoid receptor type 1 (CB1 receptor) agonist as demonstrated by binding studies and functional assays [15].

Some drugs are completely metabolized and are not excreted into urine to any appreciable extent. Should this happen to the designer amines, this would pose an additional problem as the metabolism of these drugs is generally unknown. Even if the structures of the designer amines and metabolites were known, it is difficult to obtain access to standards to relevant metabolites to these drugs. For these reasons, untargeted MS analysis provides an alternative to targeted analysis using GC or LC-MS. With the molecular formula, it may be possible to predict and identify a metabolite based on the usual means that drugs are metabolized down, i.e., through oxidation, demethylation, and conjugation.

Another important application of high resolution mass spectrometry is in detecting active and potentially illicit and dangerous drugs found in herbal medications. Since these products are considered food supplements, they are not regulated by the Food and Drug Administration as a drug. There are many formulations of herbals that contain designer drugs. Consumers take these supplements without realizing that the contents may not be natural substances but chemicals that can produce dangerous side effects. When patients suffer adverse events and are seen in an emergency department, the analysis of body fluids and the herbal medications themselves using HR-MS can be useful in managing these patients. The FDA will conduct an investigation of herbal suspect herbal medications. However, many of these products are made in Asia and list fictitious addresses.

Adulterants and contaminants is testing application that is well suited for high-resolution MS analysis. Adulterants are used in illicit drug production as a means to increase the mass of the product or to achieve a desired alternate effect. Some of these adulterants, e.g., levamisole which is widely used in cocaine, can produce toxic effects [9]. As with herbals, the analysis is conducted on both the patient's blood and urine, and the illicit drug itself, if it is available. To assist in diagnoses, physicians are reminded to ask and retain from patients and their families regarding the medications that have or are taking.

Drug testing for pain management has become a major growth area for clinical toxicology laboratories. As a condition for continued participation and reimbursement, patients with chronic pain must demonstrate adherence to their drug prescriptions, and abstain from abuse of illicit drugs. For therapeutic drug monitoring, GC-MS and LC-MS/MS are well suited for targeted quantitative testing. Some laboratories have generated urine drug testing reports that indicate compliance by the donor, through a correction for fluid volume based on urine creatinine testing. However, the variability of drug absorption, distribution, metabolism, and excretion is large, making accurate assessments impossible. For unknown analysis, MR-MS analysis is effective. The presence of an illicit drug may disqualify participants from participating in the clinic. Due to the wide menu of analytes, some pain management laboratories have eliminated immunoassay screening completely and perform all testing using mass spectrometry.

High-resolution mass spectrometry analysis offers significant advantages for clinical toxicology, and to forensic toxicology, although to a lesser extent. While quantitative analysis can be conducted by high-resolution instrumentation, tandem mass spectrometry technology is preferred. It is therefore difficult for laboratories to justify the cost of owning separate tandem and high-resolution MS equipment. An optimum situation is the acquisition of systems that have both triple-quadrupole and high-resolution detections available within a single instrument. As would be expected, the costs for a combined instrument are higher than for the individual instruments. But it enables the analyst to use the best detector suited for the application.

# 2 Materials

## 2.1 Samples

Urine, randomly collected.

## 2.2 Standards and Reagents (See Note 1)

1. Specimen preparation buffer (water containing 12.5 % 50:50 methanol:acetonitrile): Prepare 30 mL of 50:50 methanol:acetonitrile. Mix 25 mL of the 50:50 methanol:acetonitrile solution with 175 mL water to prepare 200 mL of preparation buffer. Solution is stable at 15–30 °C for 1 year.

2. 1 M Ammonium formate: Measure 6.305 g of ammonium formate using an analytical balance. Add the ammonium formate to a 100 mL volumetric flask and fill to 100 mL with water. Solution is stable at 15–30 °C for 1 year.

3. Mobile phase A (water containing 5 mM ammonium formate and 0.05 % formic acid): To prepare 2 L of mobile phase A, add 10 mL of 1 M ammonium formate, 1 L of water, and 1 mL of formic acid to a graduated cylinder, and then add water to bring the volume to 2 L. Solution is stable at 15–30 °C for 6 months.

4. Mobile phase B (50:50 methanol:acetonitrile containing 0.05 % formic acid): To prepare 2 L of mobile phase B, mix 1 L of acetonitrile with 1 L of methanol. Remove 1 mL of the mix, and then add 1 mL of formic acid. Solution is stable at 15–30 °C for 6 months.

5. Pump wash solution (50:50 water: methanol): Solution is stable at 15–30 °C for 6 months.

6. Drug-free human urine: Purchased from UTAK Laboratories (Valencia, CA).

7. Reference standards: Drug standards and deuterium-labeled internal standards were purchased from Cerilliant (Round Rock, TX), Grace Davison/Alltech (Deerfield, IL), Sigma-Aldrich (St. Louis, MO) and Lipomed Inc. (Cambridge, MA). Reference standards are stored according to manufacturer's recommendation.

8. Quality control mixes: Mixes of 20–25 drugs prepared in 50:50 methanol:acetonitrile, with each drug present at a final concentration of 1 μg/mL. Each mix contains different drugs, such that all the drugs included in the method are represented in at least one mix. Drugs are not grouped into mixes by class; instead, each mix contains drugs with varying retention times and physical/chemical properties. Mixes are stable at −20 °C for 1 year, or until the stock standard solutions expire, whichever comes first.

### 2.3 Internal Standards

1. Fentanyl-D5 Internal Standard Stock Solution (100 μg/mL) in methanol. Purchased from Cerilliant (Round Rock, TX).

2. Fentanyl-D5 Internal Standard Working Solution (1 μg/mL) in specimen preparation buffer. Add 100 μL of Fentanyl-D5 stock standard to 9.9 mL of specimen preparation buffer. Prepare 1 mL aliquots. Stable for 1 year at −20 °C, or until the stock solution expires, whichever comes first.

### 2.4 Quality Controls

1. The goal of quality control is to verify instrument performance across all drug categories and across the chromatographic separation by injection of negative control and positive control samples.

To ensure adequate control of the 200+ drugs in the method, and to reduce waste of reagents and analytical time, the positive control samples are used on a rotating schedule. The number of different positive controls depends on the total number of drugs in the method, we use ten unique controls for 200 drugs. Each day of testing a different quality control mixture is used, when we have completed 10 days of testing we will have verified all 200 drugs in our method, and the next day we begin again with mix 1.

2. Negative control: Drug-free human urine (UTAK Laboratories). Stable at 4 °C until expiration date on label.

3. Positive controls: Quality control mixes are spiked into drug free human urine at a final concentration of 100 ng/mL. Each positive control mix is stable at 4 °C for 30 days.

**2.5 Supplies**

1. Autosampler vials (2 mL) and caps with polytetrafluoroethylene (PTFE)-lined septa (Agilent Technologies, Santa Clara, CA).

2. Amber vials with PTFE-lined caps for drug and internal standard solutions (Grace Davison/Alltech, Deerfield, IL).

3. HPLC columns (Kinetex C18, 2.6 µm, 3×50 mm) and guard columns (SecurityGuard ULTRA cartridges, C18) (Phenomenex, Torrance, CA).

4. APCI calibration solution for 5600 TripleTOF® (ABSciex, Foster City, CA).

**2.6 Equipment**

1. An HPLC system compatible with ABSciex Analyst software (e.g., Shimadzu LC-20ADXR Prominence) with degasser, binary pump, solvent switching valve, temperature controlled autosampler, and temperature-controlled column compartment (Shimadzu, Pleasanton, CA).

2. Nitrogen generator capable of supplying curtain, collision, and source gases.

3. A 5600 TripleTOF® quadrupole time-of-flight mass spectrometer with a DuoSpray™ source and automatic calibrant delivery system (ABSciex, Foster City, CA).

4. Analyst® 1.5, PeakView® 2.0, and MasterView™ 1.0 software with ChemSpider plug-in (ABSciex, Foster City, CA).

# 3    Methods

**3.1 Stepwise Procedure**

1. Aliquot approximately 1 mL of each patient urine specimen using a disposable, plastic transfer pipet into a 12×75 mm plastic test tube. Cap the tube.

2. Centrifuge all patient urine specimens for 10 min at $1500 \times g$ in the centrifuge.

3. Label an amber autosampler vial for the negative control, positive control, and each patient urine. Due to the possibility of carryover, a double blank sample must be injected between each patient sample.

4. Pipet 700 μL of sample preparation buffer into each vial.

5. Pipet 100 μL of internal standard into each vial.

6. Pipet 200 μL of negative control, positive control and patient urine to the correspondingly labeled vials.

7. Pipet 1000 μL of sample preparation buffer into each double blank vial.

8. Cap the vials and vortex to mix.

9. Place the vials in the autosampler tray for testing.

10. Inject 10 μL (*see* **Note 2**).

*3.2   Instrument Operating Conditions*

1. *Liquid chromatography system*: Column and mobile phase compositions are described in Subheading 2.1.

   (a) Injection volume: 10 μL.

   (b) Flow rate: 400 μL/min.

   (c) LC parameters:

   - 0 min: 2 % MPB

   - 10 min: 98 % MPB

   - Wash for 2 min at 100 % MPB

   - Re-equilibrate for 2 min at 2 % MPB

2. 5600 QTOF mass spectrometer:

   (a) Ion source: positive electrospray, 500 °C, ion spray voltage floating 5500 V, declustering potential 100 V.

   (b) Gas settings: source gas 1—30 PSI, source gas 2—30 PSI, curtain gas—25 PSI.

   (c) Ion release delay: 67.

   (d) Ion release width: 25.

   (e) Full-scan TOF MS from 50 to 700 Da.

   (f) Information-dependent acquisition of product ion spectra for ≤20 candidate ions per cycle.

   (g) Automatic calibration verification of TOF and MS/MS mass accuracy every five injections.

*3.3   Data Analysis*

One of the rationales for collecting untargeted, full scan, HRMS data is that three different data analysis techniques can be used to maximize the information gleaned from each sample.

Targeted analysis requires the most information about each compound, but identifies compounds with very high confidence. Suspect analysis requires less information, only empirical formula, but may produce numerous false-positive results. Suspect searching is subjected to the caveat that absence of evidence is not evidence of absence. Untargeted analysis requires no a priori information, but is extremely laborious and has a low true-positive rate. All three types of data analysis described are performed using PeakView and MasterView software from ABSciex.

1. Targeted analysis: To perform targeted analysis, the analyst must know the accurate mass, chromatographic retention time, isotope pattern, and product ion spectrum of the targeted compounds. Accurate mass and the expected isotope pattern can be predicted based on the empirical formula of the compound of interest. Retention times must be established and product ion spectra must be collected on the system where the method will be run. Product ion spectra can be collected using a dedicated product ion scan and added to the compound library using Library View (ABSciex). Retention times are established by averaging the observed retention time from ≥3 injections on ≥2 individual columns. Targeted methods may include any number of compounds. An XIC list is built in MasterView using a precursor mass search of ±30 ppm, a retention time window of ±15 s, and a minimum peak intensity of 1 count per second. Positivity is assessed using combined scores. Any compound with a combined score ≥70, that passes visual inspection, is considered positive. The combined score is a weighted average of the following parameters: 10 % mass error, 10 % retention time error, 10 % isotope error, and 70 % library match.

2. Suspect analysis: The analyst only needs to know the accurate mass and expected isotope pattern for each compound in the XIC list in order to perform a suspect screen. Retention times and product ion spectra are not required. Using a meta-library of product ion spectra does increase the fidelity of the suspect search, but it is not essential. XIC lists can contain many thousands of compounds. Results from a suspect search can be used to identify possible hits, but a reference standard must be ordered and the hit must be confirmed with retention time and product ion spectral data. If library searching is enabled: compounds with a combined score ≥78 are considered preliminary positive. Combined score is based on 10 % mass error, 20 % isotope error, and 70 % library match. If library searching is not enabled: compounds with a combined score of ≥76 are considered preliminary positive. Combined score is based on 40 % mass error and 60 % isotope error.

3. Untargeted analysis: Untargeted analysis requires no a priori information on compounds present in the sample. Instead of telling the software what masses to look for, the software provides the user with a list of masses, product ion spectra, and peak intensities for ions in the sample. It is then up to the analyst to identify the relevant masses, an entirely empirical process. Once the masses are selected, the software uses the accurate mass and the isotope pattern to predict the molecular formula of the peak of interest. Typically, several possible formulae are presented to the user, who has to select the correct one. While some possible formulae are obviously impossible, this step is also largely empirical and may require investigating several formulae. After selecting the correct formula, the user can transfer the information into the built in ChemSpider interface to attempt to determine the specific structural isomer that is present in the sample. The ChemSpider plug-in can be configured to use fragmentation information to help select the correct structural isomer. Untargeted analysis can be used to identify tentative hits, but like suspect screening, a reference standard must be used to confirm that the retention time and product ion spectrum match the compound in the sample. For most purposes, suspect analysis is a faster and more robust method of searching samples for an extended list of compounds.

## 4 Notes

1. All reagents, including water and solvents, must be analytical grade or better. All glassware used to prepare reagents should be free of detergents or other residue. For measurement purposes, Class A volumetric glassware should be used.

2. Due to the possibility of carryover, a double-blank sample must be injected between each patient sample.

## References

1. Renagarajan A, Mullins ME (2013) How often do false-positive phencyclidine urine screens occur with use of common medications? Clin Toxicol (Phila) 51(6):493–496
2. Wenk RE (2006) False-negative urine immunoassay after lorazepam overdose. Arch Pathol Lab Med 130:1600–1601
3. Wu AHB, Gerona R, Armenian P, French D, Petrie M, Lynch KL (2012) Role of liquid chromatography-high-resolution mass spectrometry (LC-HR/MS) in clinical toxicology. Clin Toxicol 50:733–742
4. Ojanpera I, Kolmonen M, Pelander A (2012) Current use of high-resolution mass spectrometry in drug screening relevant to clinical and forensic toxicology and doping control. Anal Bioanal Chem 403:1203–1220
5. Crews BO, Pesce AJ, West R, Nguye H, Fitzgerald RL (2012) Evaluation of high-resolution mass spectrometry for urine toxicology screening in a pain management setting. J Anal Toxicol 36:601–607
6. Croley TR, White KD, Callahan JH, Musser SM (2012) The chromatographic role in high

resolution mass spectrometry for non-targeted analysis. J Am Soc Mass Spectrom 23:1569–1578

7. Bush DM (2008) The U.S. Mandatory Guidelines for Federal Workplace Drug Testing Programs: current status and future considerations. Forensic Sci Int 174:111–119

8. Jarvie DR, Simpson D (1986) Drug screening: evaluation of the Toxi-lab TLC system. Ann Clin Biochem 23:76–84

9. Demedts P, Wauters A, Franck F, Neels H (1994) Evaluation of the Remedi drug profiling system. Eur J Clin Chem Clin Biochem 32:409–417

10. Gerona RR, Wu AH (2012) Bath salts. Clin Lab Med 32(3):415–427

11. Fattore L, Fratta W (2011) Beyond THC: the new generation of cannabinoid designer drugs. Front Behav Neurosci 5(60):1–12

12. Stogner JM (2014) The potential threat of acetyl fentanyl: legal issues, contaminated heroin, and acetyl fentanyl "disguised" as other opioids. Ann Emerg Med 64(6):637–639

13. Petrie MS, Lynch KL, Wu AHB, Steinhardt AA, Horowitz GL (2012) Prescription compliance or illicit designer drug abuse? Clin Chem 58:1631–1635

14. Graf J, Lynch K, Yeh CL, Tarter L, Richman N, Nguyen T, Kral A, Dominy S, Imboden J (2011) Purpura, cutaneous necrosis, and anti-neutrophil cytoplasmic antibodies associated with levamisole-adulterated cocaine. Arthritis Rheum 63:3998–4001

15. Title 21 US Code (USC) Controlled Substance Act. US Department of Justics, Drug Enforcement Agency. Office of Diversion Control. http://www.deadiversion.ucdoj.gov/21cfr/21usc/812.htm

# Chapter 18

# Quantitation of Ethyl Glucuronide and Ethyl Sulfate in Urine Using Liquid Chromatography-Tandem Mass Spectrometry (LC-MS/MS)

## Matthew H. Slawson and Kamisha L. Johnson-Davis

### Abstract

Ethyl glucuronide and ethyl sulfate are minor conjugated metabolites of ethanol that can be detected in urine for several days after last ingestion of ethanol. The monitoring of ethanol use has both clinical and forensic applications and a longer detection window afforded by monitoring these metabolites is obvious. LC-MS/MS is used to analyze diluted urine with deuterated analogs of each analyte as internal standards to ensure accurate quantitation and control for any potential matrix effects. High aqueous HPLC is used to chromatograph the metabolites. Negative ion electrospray is used to introduce the metabolites into the mass spectrometer. Selected reaction monitoring of two product ions for each analyte allows for the calculation of ion ratios which ensures correct identification of each metabolite, while a matrix-matched calibration curve is used for quantitation.

**Key words** Ethyl glucuronide, Ethyl sulfate, Urine, Mass spectrometry

## 1   Introduction

Ethyl glucuronide and ethyl sulfate are minor phase II metabolites of ethanol formed by enzymatic conjugation of ethanol with either glucuronic or sulfonic acid [1]. Ethanol is normally detected in urine for only a few hours after last ingestion [1], and ethanol can be produced under certain circumstances in vitro due to fermentation of urine samples [2–7]. Ethyl glucuronide and ethyl sulfate however, can be detected up to several days after last ingestion [8–17]. In recent years, ethyl glucuronide and ethyl sulfate have become useful diagnostic biomarkers for determining recent ethanol use in abstinence or other treatment programs [8, 13, 14, 18–29]. Because ethyl glucuronide and ethyl sulfate are the products of biotransformation, their measurement can also be used to eliminate the misidentification of alcohol use due to in vitro fermentation processes [2–7]. Also, recent studies indicate the presence of

Uttam Garg (ed.), *Clinical Applications of Mass Spectrometry in Drug Analysis: Methods and Protocols*, Methods in Molecular Biology, vol. 1383, DOI 10.1007/978-1-4939-3252-8_18, © Springer Science+Business Media New York 2016

ethyl glucuronide in urine of people exposed to alcohol containing hygiene products such as mouthwash or hand sanitizer. The concomitant presence of ethyl sulfate in the urine has been suggested in minimizing this potential misinterpretation [30–32].

## 2  Materials

### 2.1  Samples

Random urine collection. Specimens can be stored for at least 20 days (refrigerated or frozen) prior to analysis.

### 2.2  Reagents

1. Mobile Phase A: 0.1 % formic acid in deionized water.

2. Mobile Phase B: 0.1 % formic acid in acetonitrile.

3. Certified negative urine. Synthetic urine is NOT recommended (*see* **Note 1**).

### 2.3  Standards and Calibrators

1. Ethyl-β-D-glucuronide, 1.0 mg/mL in methanol (Cerilliant, Round Rock, TX).

2. Ethyl sulfate sodium salt, 1.0 mg/mL (as ethyl sulfate) stock in methanol (Cerilliant, Round Rock, TX).

3. Intermediate working standard solution containing ethyl glucuronide and ethyl sulfate at 100 ng/μL in methanol:water (1:1): Add ~3 mL of methanol:water (1:1) to a 5 mL volumetric flask. Add 0.5 mL of each calibrator reference material to the flask, QS to 5 mL with methanol:water (1:1), add a stir bar and stopper and mix for 30 min at room temperature. Aliquot as appropriate for subsequent use. Store frozen, stable for 1 year. This volume can be scaled up or down as appropriate.

4. Working calibrators: Add approximately 5 mL certified negative urine to a labeled volumetric flask. Add the appropriate volume/concentration as shown in Table 1 of standard mate-

**Table 1**
**Preparation of calibrators**

| Final concentration ng/mL | Solution μL | Stock solution concentration |
|---|---|---|
| 10,000 | 100 | 1 mg/mL |
| 7500 | 750 | 100 ng/μL |
| 5000 | 500 | 100 ng/μL |
| 1000 | 100 | 100 ng/μL |
| 500 | 50 | 100 ng/μL |
| 100 | 10 | 100 ng/μL |

For each concentration, the total volume is made to 10 mL with drug-free human urine

rial to the flask. QS to 10 mL using certified negative urine. Add a stir bar and stopper and mix for 30 min at room temperature. Aliquot as appropriate for future use. Store aliquots frozen, stable for 1 year. This volume can be scaled up or down as appropriate (*see* **Note 2**).

**2.4 Controls and Internal Standard**

1. Controls may be purchased from a third party (e.g. UTAK, Valencia, CA) and prepared according to the manufacturer. They can also prepared in-house independently from calibrators' source material using Table 1 as a guideline (*see* **Note 2**).

2. Internal Standard: Ethyl-β-D-glucuronide-D$_5$ 1.0 mg/mL in methanl frozen, stable for 1 year (*see* **Note 2**).

**2.5 Supplies**

1. Transfer/aliquoting pipettes and tips.

2. Instrument compatible autosampler vials with injector appropriate caps or 96 deepwell plate and capping mat.

3. Synergi 2.5 μm Hydro-RP 100A HPLC column, size 100×2 mm (Phenomenex, Torrance, CA).

4. Security Guard Kit with Polar RP Security Guard Cartridges, 4.0×2.0 mm ID (Phenomenex, Torrance, CA).

5. Compatible centrifuge tubes (if needed, see below). These are to be used to centrifuge urine specimens assuming the collection containers are not compatible with a high speed centrifuge.

**2.6 Equipment**

1. High speed centrifuge capable of holding patient sample tubes or other compatible centrifuge tubes.

2. Vortex mixer.

3. Low dead volume binary HPLC pump, thermostatted column compartment with switching valve, vacuum degasser (Agilent, Santa Clara, CA), and DLW CTC PAL auto-sampler (Leap Technologies, Carrboro, NC).

4. API 4000 LC-MS/MS system with Turbo Ion source running current version of Analyst software (ABSciex, Framingham, MA).

# 3 Methods

**3.1 Stepwise Procedure**

1. If necessary, transfer a representative portion of the patient sample to a compatible centrifuge tube. Centrifuge patient samples at ≥1000×$g$ for 10 min to remove any particulates.

2. Aliquot 50 μL of each clarified patient sample, calibrator and QC into appropriately labeled autosampler vial or 96-well plate.

3. Add 10 μL of internal standard to each vial or well.

4. Add 500 μL of fresh CLRW to each vial or well.

5. Cap each vial or plate and vortex briefly.

6. Analyze on LC-MS/MS.

**3.2 Instrument's Operating Conditions**

1. Table 2 summarizes typical LC conditions.

2. Table 3 summarizes typical MS conditions.

3. Table 4 summarizes typical MRM conditions.

4. Each instrument should be individually optimized for best method performance (*see* **Notes 3–6**).

**3.3 Data Analysis**

1. Representative MRM chromatograms of ethyl glucuronide and ethyl sulfate are shown in Fig. 1 (*see* **Notes 3** and **7**).

2. The dynamic range for this assay is 100–10,000 ng/mL for each analyte. Samples exceeding this range can be diluted 5× or 10× as needed to achieve an accurate calculated concentration, if needed (*see* **Note 1**).

**Table 2**
**Typical HPLC conditions**

| Cycle name | Analyst LC-Inj DLW standard_Rev05 | | | |
|---|---|---|---|---|
| Wash solvent 1 | CLRW | | | |
| Wash solvent 2 | Methanol | | | |
| Injection volume | 5 L | | | |
| Vacuum degassing | On | | | |
| Temperature | 30 °C | | | |
| A reservoir | 0.1 % HCOOH in CLRW | | | |
| B reservoir | 100 % acetonitrile | | | |
| Gradient table | | | | |
| Step | Time (min) | Flow (μL/min) | A (%) | B (%) |
| 0 | 0 | 300 | 100 | 0 |
| 1 | 1 | 300 | 100 | 0 |
| 2 | 2.5 | 300 | 85 | 15 |
| 3 | 3.25 | 300 | 55 | 45 |
| 4 | 3.3 | 300 | 10 | 90 |
| 5 | 3.8 | 350 | 10 | 90 |
| 6 | 4 | 350 | 100 | 0 |
| 7 | 5 | 300 | 100 | 0 |

**Table 3**
**Typical mass spectrometer conditions**

| Parameter | Value |
|---|---|
| CUR | 10 |
| GS1 | 50 |
| GS2 | 50 |
| TEM | 700 |
| ihe | ON |
| CAD | 4 |
| IS | −4500 |
| EP | −10 |
| Scan type | MRM |
| Scheduled MRM | No |
| Polarity | Negative |
| Scan mode | N/A |
| Ion source | Turbo spray |
| Resolution Q1 | Unit |
| Resolution Q3 | Low |
| Intensity thres. | 0.00 cps |
| Settling time | 0.0000 ms |
| MR pause | 5.0070 ms |
| MCA | No |
| Step size | 0.00 Da |
| Dwell (ms) | 100 |

3. Data analysis is performed using the Analyst software to integrate peaks, calculate peak area ratios, and construct calibration curves using a linear 1/x weighted fit ignoring the origin as a data point. Sample concentrations are then calculated using the derived calibration curves (*see* **Note 2**).

4. Calibration curves should have $r^2$ value $\geq 0.99$.

5. Typical imprecision is <15 % both inter- and intra-assay.

6. An analytical batch is considered acceptable if chromatography is acceptable (*see* **Note 3**) and QC samples calculate to within 20 % if their target values and ion ratios are within 20 % of the calibrator ion ratios (*see* **Note 1**).

**Table 4**
**Typical MRM conditions**

| Analyte | Q1 Mass (Da) | Q3 Mass (Da) | Param |
|---------|--------------|--------------|-------|
| Ethyl sulfate (qualifier) | 124.8 | 79.8 | DP = –38.00<br>CE = –36.00<br>CXP = –6.50 |
| | Approx. retention time | 1.4 min | |
| Ethyl sulfate (quant.) | 124.8 | 96.8 | DP = –38.00<br>CE = –22.00<br>CXP = –6.50 |
| | Approx. retention time | 1.4 min | |
| Ethyl sulfate-D5 (internal standard) | 130.1 | 98.1 | DP = –38.00<br>CE = –22.00<br>CXP = –6.50 |
| | Approx. retention time | 1.4 min | |
| Ethyl glucuronide (qualifier) | 220.6 | 74.8 | DP = –44.00<br>CE = –22.00<br>CXP = –5.00 |
| | Approx. retention time | 2.5 min | |
| Ethyl glucuronide (quant.) | 220.6 | 84.9 | DP = –44.00<br>CE = –22.00<br>CXP = –5.00 |
| | Approx. retention time | 2.5 min | |
| Ethyl gluc.-D5 (internal standard) | 226.1 | 85.1 | DP = –44.00<br>CE = –22.00<br>CXP = –5.00 |
| | Approx. retention time | 2.4 min | |

## 4    Notes

1. Ion suppression can occur in human urine at the retention time window for ethyl glucuronide. This ion suppression does not occur in synthetic urine, processed urine (i.e., filtered, frozen, diluted, etc.) or water alone. It is therefore important that calibrators and QCs be prepared in unprocessed human urine to ensure an equivalent instrument response between control and patient samples. The suppression affects the analyte and I.S. equally and is fairly consistent among typical urine specimens. Ion suppression should be effectively controlled by the labeled internal standard; however, internal standard signal should be reviewed in each specimen and any significant deviation in area counts should be investigated for excessive ion suppression. Sample dilution may be appropriate to compensate for excessive ion suppression, but may compromise the ability to report

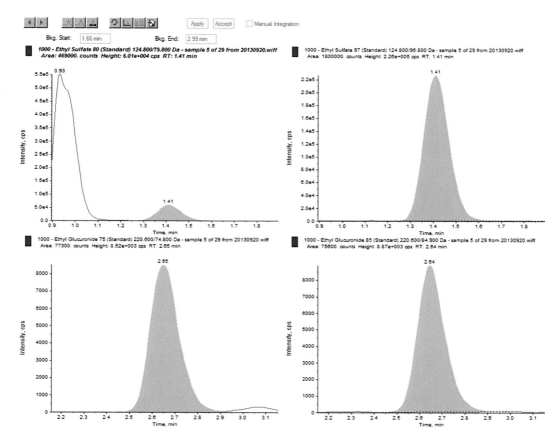

**Fig. 1** Typical MRM chromatogram for ethyl glucuronide and ethyl sulfate. 1000 ng/mL diluted from fortified human urine and analyzed according to the described method

low concentrations of analyte. Ion suppression may also occur due to extremely high concentrations of ethyl glucuronide in the specimen; in this case sample dilution may be appropriate for an accurate concentration to be determined.

2. Validate/verify all calibrators, QCs, internal standard before placing into use.

3. A large endogenous peak (*see* Fig. 1) typically elutes in human urine specimens in the MRM window for the ethyl sulfate qualifier ion (124.8 → 79.8). This can be controlled with the use of a divert valve or if more chromatographic separation is required, a longer column may be employed at the cost of a longer run cycle time.

4. This method uses negative ion electrospray; excessive arcing at the ESI needle tip is possible and can compromise the performance of this assay and therefore must be minimized:

   (a) Ensure the ESI needle tip does not protrude too far out the end of the probe.

(b) Maintain the spray voltage as low as possible without compromising ion signal.

(c) Utilize zero grade air in the nebulizer instead of pure Nitrogen.

5. Maintain cleanliness of the source by maintaining a flow of curtain gas that does not compromise method performance.

6. HPLC conditions for this method utilize high aqueous to achieve retention of ethyl sulfate on the LC column. It is important to minimize the amount of organic solvent that the sample contacts prior to introduction into the instrument or retention times and peak shapes will be compromised. For example:

(a) Excessive organic solvent in the sample prep.

(b) Strong solvent as the final wash cycle of the autosampler. Ensure that strong wash solvent (>1 % organic) does not remain in the injector path as a result of the DLW wash cycle. Use 100 % aqueous in wash reservoir#1 as suggested in the CTC DLW documentation.

7. The response for ethyl glucuronide is 1–2 orders of magnitude less than ethyl sulfate at an equivalent concentration.

## References

1. Baselt RC (2011) Disposition of toxic drugs and chemicals in man, 9th edn. Biomedical Publications, Seal Beach, CA

2. Gruszecki AC, Robinson CA, Kloda S, Brissie RM (2005) High urine ethanol and negative blood and vitreous ethanol in a diabetic woman: a case report, retrospective case survey, and review of the literature. Am J Forensic Med Pathol 26:96–98

3. Jones AW, Eklund A, Helander A (2000) Misleading results of ethanol analysis in urine specimens from rape victims suffering from diabetes. J Clin Forensic Med 7:144–146

4. Kugelberg FC, Jones AW (2007) Interpreting results of ethanol analysis in postmortem specimens: a review of the literature. Forensic Sci Int 165:10–29

5. Saady JJ, Poklis A, Dalton HP (1993) Production of urinary ethanol after sample collection. J Forensic Sci 38:1467–1471

6. Singer PP, Jones GR (1997) Very unusual ethanol distribution in a fatality. J Anal Toxicol 21:506–508

7. Sulkowski HA, Wu AH, McCarter YS (1995) In-vitro production of ethanol in urine by fermentation. J Forensic Sci 40:990–993

8. Albermann ME, Musshoff F, Doberentz E, Heese P, Banger M, Madea B (2012) Preliminary investigations on ethyl glucuronide and ethyl sulfate cutoffs for detecting alcohol consumption on the basis of an ingestion experiment and on data from withdrawal treatment. Int J Leg Med 126:757–764

9. Albermann ME, Musshoff F, Madea B (2012) A high-performance liquid chromatographic-tandem mass spectrometric method for the determination of ethyl glucuronide and ethyl sulfate in urine validated according to forensic guidelines. J Chromatogr Sci 50:51–56

10. Beyer J, Vo TN, Gerostamoulos D, Drummer OH (2011) Validated method for the determination of ethylglucuronide and ethylsulfate in human urine. Anal Bioanal Chem 400:189–196

11. Hegstad S, Johnsen L, Morland J, Christophersen AS (2009) Determination of ethylglucuronide in oral fluid by ultra-performance liquid chromatography- tandem mass spectrometry. J Anal Toxicol 33:204–207

12. Kissack JC, Bishop J, Roper AL (2008) Ethylglucuronide as a biomarker for ethanol detection. Pharmacotherapy 28:769–781

13. Kummer N, Wille S, Di Fazio V, Lambert W, Samyn N (2013) A fully validated method for the quantification of ethyl glucuronide and

ethyl sulphate in urine by UPLC-ESI-MS/MS applied in a prospective alcohol self-monitoring study. J Chromatogr B Analyt Technol Biomed Life Sci 929:149–154

14. Niemela O (2007) Biomarkers in alcoholism. Clin Chim Acta 377:39–49

15. Politi L, Morini L, Groppi A, Poloni V, Pozzi F, Polettini A (2005) Direct determination of the ethanol metabolites ethyl glucuronide and ethyl sulfate in urine by liquid chromatography/electrospray tandem mass spectrometry. Rapid Commun Mass Spectrom 19:1321–1331

16. Rosano TG, Lin J (2008) Ethyl glucuronide excretion in humans following oral administration of and dermal exposure to ethanol. J Anal Toxicol 32:594–600

17. Weinmann W, Schaefer P, Thierauf A, Schreiber A, Wurst FM (2004) Confirmatory analysis of ethylglucuronide in urine by liquid-chromatography/electrospray ionization/tandem mass spectrometry according to forensic guidelines. J Am Soc Mass Spectrom 15:188–193

18. Bendroth P, Kronstrand R, Helander A, Greby J, Stephanson N, Krantz P (2008) Comparison of ethyl glucuronide in hair with phosphatidylethanol in whole blood as post-mortem markers of alcohol abuse. Forensic Sci Int 176:76–81

19. Concheiro M, Cruz A, Mon M, de Castro A, Quintela O, Lorenzo A, Lopez-Rivadulla M (2009) Ethylglucuronide determination in urine and hair from alcohol withdrawal patients. J Anal Toxicol 33:155–161

20. Dufaux B, Agius R, Nadulski T, Kahl HG (2012) Comparison of urine and hair testing for drugs of abuse in the control of abstinence in driver's license re-granting. Drug Test Anal 4:415–419

21. Grosse J, Anielski P, Sachs H, Thieme D (2009) Ethylglucuronide as a potential marker for alcohol-induced elevation of urinary testosterone/epitestosterone ratios. Drug Test Anal 1:526–530

22. Helander A, Peter O, Zheng Y (2012) Monitoring of the alcohol biomarkers PEth, CDT and EtG/EtS in an outpatient treatment setting. Alcohol Alcohol 47:552–557

23. Marques PR (2012) Levels and types of alcohol biomarkers in DUI and clinic samples for estimating workplace alcohol problems. Drug Test Anal 4:76–82

24. Morini L, Marchei E, Tarani L, Trivelli M, Rapisardi G, Elicio MR, Ramis J, Garcia-Algar O, Memo L, Pacifici R et al (2013) Testing ethylglucuronide in maternal hair and nails for the assessment of fetal exposure to alcohol: comparison with meconium testing. Ther Drug Monit 35:402–407

25. Mutschler J, Grosshans M, Koopmann A, Mann K, Kiefer F, Hermann D (2010) Urinary ethylglucuronide assessment in patients treated with disulfiram: a tool to improve verification of abstention and safety. Clin Neuropharmacol 33:285–287

26. Stewart SH, Koch DG, Burgess DM, Willner IR, Reuben A (2013) Sensitivity and specificity of urinary ethyl glucuronide and ethyl sulfate in liver disease patients. Alcohol Clin Exp Res 37:150–155

27. Thieme D, Grosse J, Keller L, Graw M (2011) Urinary concentrations of ethyl glucuronide and ethyl sulfate as thresholds to determine potential ethanol-induced alteration of steroid profiles. Drug Test Anal 3:851–856

28. Turfus SC, Vo T, Niehaus N, Gerostamoulos D, Beyer J (2013) An evaluation of the DRI-ETG EIA method for the determination of ethyl glucuronide concentrations in clinical and post-mortem urine. Drug Test Anal 5:439–445

29. Winkler M, Skopp G, Alt A, Miltner E, Jochum T, Daenhardt C, Sporkert F, Gnann H, Weinmann W, Thierauf A (2013) Comparison of direct and indirect alcohol markers with PEth in blood and urine in alcohol dependent inpatients during detoxication. Int J Legal Med 127:761–768

30. Pragst F, Spiegel K, Sporkert F, Bohnenkamp M (2000) Are there possibilities for the detection of chronically elevated alcohol consumption by hair analysis? A report about the state of investigation. Forensic Sci Int 107:201–223

31. Reisfield GM, Goldberger BA, Crews BO, Pesce AJ, Wilson GR, Teitelbaum SA, Bertholf RL (2011) Ethyl glucuronide, ethyl sulfate, and ethanol in urine after sustained exposure to an ethanol-based hand sanitizer. J Anal Toxicol 35:85–91

32. Hoiseth G, Yttredal B, Karinen R, Gjerde H, Christophersen A (2010) Levels of ethyl glucuronide and ethyl sulfate in oral fluid, blood, and urine after use of mouthwash and ingestion of nonalcoholic wine. J Anal Toxicol 34:84–88

# Chapter 19

# Quantification of Hydroxychloroquine in Blood Using Turbulent Flow Liquid Chromatography-Tandem Mass Spectrometry (TFLC-MS/MS)

## Allison B. Chambliss, Anna K. Füzéry, and William A. Clarke

## Abstract

Hydroxychloroquine (HQ) is used routinely in the treatment of autoimmune disorders such as rheumatoid arthritis and lupus erythematosus. Issues such as marked pharmacokinetic variability and patient non-compliance make therapeutic drug monitoring of HQ a useful tool for management of patients taking this drug. Quantitative measurements of HQ may aid in identifying poor efficacy as well as provide reliable information to distinguish patient non-compliance from refractory disease. We describe a rapid 7-min assay for the accurate and precise measurement of HQ concentrations in 100 µL samples of human blood using turbulent flow liquid chromatography coupled to tandem mass spectrometry. HQ is isolated from EDTA whole blood after a simple extraction with its deuterated analog, hydroxychloroquine-d4, in 0.33 M perchloric acid. Samples are then centrifuged and injected onto the TFLC-MS/MS system. Quantification is performed using a nine-point calibration curve that is linear over a wide range (15.7–4000 ng/mL) with precisions of <5 %.

**Key words** Hydroxychloroquine, Therapeutic drug monitoring, Turbulent flow liquid chromatography, Tandem mass spectrometry, Blood, Quantification

## 1 Introduction

Hydroxychloroquine (HQ), 2-[4-[(7-chloroquinolin-4-yl)amino]pentyl-ethylamino]ethanol, is a hydroxylated form of chloroquine, an aminoquinoline first synthesized in the 1930s for the treatment of malaria. Since then, HQ has also shown effectiveness in the treatment of several autoimmune and inflammatory disorders including lupus erythematosus, rheumatoid arthritis, chronic Q fever, Sjögren's syndrome, and various skin diseases [1].

HQ is commonly prescribed in oral doses of 200 or 400 mg per day. However, HQ is characterized by a long delay in onset of action and demonstrates wide pharmacokinetic variability among patients. Elimination half-life is estimated at up to 40 days [2]. Several studies indicate that low blood HQ concentrations may

Uttam Garg (ed.), *Clinical Applications of Mass Spectrometry in Drug Analysis*: Methods and Protocols, Methods in Molecular Biology, vol. 1383, DOI 10.1007/978-1-4939-3252-8_19, © Springer Science+Business Media New York 2016

predict disease exacerbation [3, 4]. Additionally, measurement of blood HQ concentrations can identify patients considered non-compliant to treatment regimens and may improve management of refractory disease [5].

Previous methods of quantifying HQ blood levels, which included HPLC with fluorescence detection, consisted of extensive preparatory steps and run times of at least 15 min [6, 7]. Liquid chromatography paired with mass spectrometry may offer a more efficient method for HQ analysis [8]. The following chapter describes a simple, rapid (7 min), accurate and precise method to measure HQ in whole blood samples using turbulent flow liquid chromatography coupled to electrospray-positive ionization tandem mass spectrometry (TFLC-MS/MS) [9]. In this assay, HQ is isolated from 100 μL of EDTA whole blood after a simple extraction with an internal standard, hydroxychloroquine-d4, in perchloric acid solution. Samples are then centrifuged and ready to be injected onto the TFLC-MS/MS system.

## 2  Materials

### 2.1  Samples

Whole blood samples collected by standard venipuncture are required for analysis. Whole blood should be collected in EDTA tubes. Prior to analysis, specimens should be stored at 4 °C and analyzed within 2 weeks. Specimens may be stored at –20 °C for up to 6 months.

### 2.2  Solvents and Reagents

1. Perchloric acid, 70 % solution in water.

2. Mobile Phase A (10 mM ammonium formate and 0.1 % formic acid in water): Add 0.7 mL of ammonium hydroxide stock and 1.4 mL of formic acid to a 1 L volumetric flask filled with 990 mL of water. Bring to full volume with water. Stable at 25 °C for up to 1 month.

3. Mobile Phase B (10 mM ammonium formate and 0.1 % formic acid in methanol): Add 0.7 mL of ammonium hydroxide stock and 1.4 mL of formic acid to a 1 L volumetric flask filled with 990 mL of methanol. Bring to full volume with methanol. Stable at 25 °C for up to 1 month.

4. Mobile Phase C (40:40:20 isopropanol, acetonitrile, acetone): Fill a 4 L screw-cap glass bottle with 1.6 L of isopropanol. Add 1.6 L of acetonitrile and 800 mL of acetone. Close bottle and invert, allow to vent, then close and store. Stable at 25 °C for up to 1 week.

5. 0.33 M perchloric acid solution: Add 14.27 mL of 70 % perchloric acid to a 500 mL volumetric flask. Fill to full volume with water and invert to mix. Decant into glass bottle. Stable at 4 °C for up to 6 months.

6. Human drug-free pooled EDTA (lavender top) blood.

**2.3  Internal Standards and Standards**

1. Primary standard: Hydroxychloroquine sulfate.
2. Primary internal standard: Hydroxychloroquine-d4 sulfate.
3. Internal standard stock solution: Add 1 mL of water to 1 mg of hydroxychloroquine-d4 to make a 1 mg/mL solution. Stable at 4 °C for up to 6 months.
4. Internal Standard Working Solution: Add 10 µL of the 1 mg/mL hydroxychloroquine-d4 stock to 499.990 mL 0.33 M perchloric acid to make a 20 ng/mL solution. Stable at 4 °C for up to 6 months.

**2.4  Calibrators and Controls**

1. Calibrators: Prepare Calibrator 9 at 4000 ng/mL by spiking 80 µL of a 1 mg/mL solution of hydroxychloroquine into 20 mL of EDTA blood. Prepare Calibrators 1–8 by making serial dilutions of Calibrator 9 as described in Table 1. For each dilution step, add 10 mL of the previous spiked calibrator with 10 mL of drug-free blood and mix. The calibrators should be aliquoted to 500 µL and are stable at –80 °C for up to 1 year.
2. Controls: As described in Table 2, prepare a 1500 ng/mL "high" control by spiking 30 µL of a 1 mg/mL solution of hydroxychloroquine into 20 mL of EDTA blood. Prepare a 50 ng/mL "low" control by spiking 1 µL of a 1 mg/mL solution of hydroxychloroquine into 20 mL of EDTA blood. The controls should be aliquoted to 500 µL and are stable at –80 °C for up to 1 year.

**Table 1**
**Preparation of hydroxychloroquine calibrators**

| Calibrator | Volume of previous standard (mL) | Drug-free whole blood (mL) | Final concentration (ng/mL) |
|---|---|---|---|
| 9 | 0.08 (1 mg/mL stock solution) | 20 | 4000 |
| 8 | 10 (#9) | 10 | 2000 |
| 7 | 10 (#8) | 10 | 1000 |
| 6 | 10 (#7) | 10 | 500 |
| 5 | 10 (#6) | 10 | 250 |
| 4 | 10 (#5) | 10 | 125 |
| 3 | 10 (#4) | 10 | 62.5 |
| 2 | 10 (#3) | 10 | 31.25 |
| 1 | 10 (#2) | 10 | 15.65 |

**Table 2**
**Preparation of hydroxychloroquine quality controls**

| Control | Volume of 1 mg/mL stock solution (μL) | Drug-free whole blood (mL) | Final concentration (ng/mL) |
|---|---|---|---|
| High | 30 | 20 | 1500 |
| Low | 1 | 20 | 50 |

**Table 3**
**HPLC gradient for detection of hydroxychloroquine**

| Step | Start time (min) | Duration (s) | TFLC system Cyclone 50 × 0.5 mm Flow (mL/min) | Grad | %A[a] | %B[b] | %C[c] | TEE | Loop | LX system HypersilGold C8 50 × 2.1 mm, 70 °C Flow (mL/min) | Grad | %A[a] | %B[b] |
|---|---|---|---|---|---|---|---|---|---|---|---|---|---|
| 1 | 0 | 30 | 1.50 | Step | 100.0 | – | – | – | Out | 0.70 | Step | 100.0 | – |
| 2 | 0.50 | 30 | 0.20 | Step | 100.0 | – | – | TEE | In | 0.70 | Step | 100.0 | – |
| 3 | 1.00 | 30 | 1.00 | Step | – | – | 100.00 | – | In | 0.70 | Ramp | 50.0 | 50.0 |
| 4 | 1.50 | 30 | 1.00 | Step | – | 100.0 | – | – | In | 0.70 | Ramp | 25.0 | 75.0 |
| 5 | 2.00 | 30 | 1.00 | Step | – | – | 100.00 | – | In | 0.70 | Ramp | – | 100.0 |
| 6 | 2.50 | 30 | 1.00 | Step | – | 100.0 | – | – | In | 0.70 | Ramp | 100.0 | – |
| 7 | 3.00 | 30 | 0.50 | Step | 100.0 | – | – | – | In | 0.70 | Ramp | – | 100.0 |
| 8 | 3.50 | 45 | 1.00 | Step | 65.0 | 35.0 | – | – | In | 0.70 | Step | – | 100.0 |
| 9 | 4.25 | 45 | 1.50 | Step | 100.0 | – | – | – | Out | 0.70 | Step | 100.0 | – |

[a]Mobile Phase A: 10 mM ammonium formate and 0.1 % formic acid in water
[b]Mobile Phase B: 10 mM ammonium formate and 0.1 % formic acid in methanol
[c]Mobile Phase C: 40:40:20 isopropanol, acetonitrile, acetone

**2.5 Analytical Equipment and Supplies**

1. ThermoFisher TSQ Vantage tandem mass spectrometer with HESI probe.

2. TFLC column: Cyclone, 50 × 0.5 mm.

3. Analytical column: Hypersil Gold C8, 3 μm, 50 × 2.1 mm.

4. 1.5 mL microcentrifuge tubes.

5. Assign instrumental operating parameters according to Tables 3 and 4. Parameters are instrumentation-specific.

**Table 4**
**TSQ Vantage instrument parameters and ion transitions**

| | |
|---|---|
| Spray voltage (V) | 4000 |
| Sheath gas pressure | 40 |
| Ion sweep gas pressure | 2.0 |
| Aux gas pressure | 20 |
| Capillary temperature (°C) | 200 |
| *Hydroxychloroquine* | |
| Parent ion | 336.2 |
| Product ion | 247.1 |
| Collision energy | 20 |
| S-lens RF amplitude | 133 |
| *Hydroxychloroquine-d4* | |
| Parent ion | 340.2 |
| Product ion | 251.1 |
| Collision energy | 21 |
| S-lens RF amplitude | 137 |

# 3 Methods

### 3.1 Stepwise Procedure

1. Pipet 100 μL of whole blood to a 1.5 mL microcentrifuge tube.

2. Add 1000 μL of Internal Standard Working Solution to the blood.

3. Cap and vortex the mixture for 30 s.

4. Centrifuge the sample at $13,400 \times g$ for 5 min.

5. Transfer the supernatant to a glass vial for loading into the autosampler.

6. Inject 10 μL of sample onto the TFLC-MS/MS system.

7. Instrumental operating parameters are given in Tables 3 and 4. Include a 2 min wash step between samples (*see* **Note 1**).

### 3.2 Analysis

1. Data are analyzed using Thermo Scientific TraceFinder software (*see* **Note 2**). The program uses the values obtained for the calibrators to construct a calibration curve based on the ratio of each calibrator to the internal standard. This calibration curve is then used to quantitate the unknowns.

2. Liquid chromatography retention times for hydroxychloroquine and hydroxychloroquine-d4 are set at $2.08 \pm 0.4$ min (*see* Fig. 1).

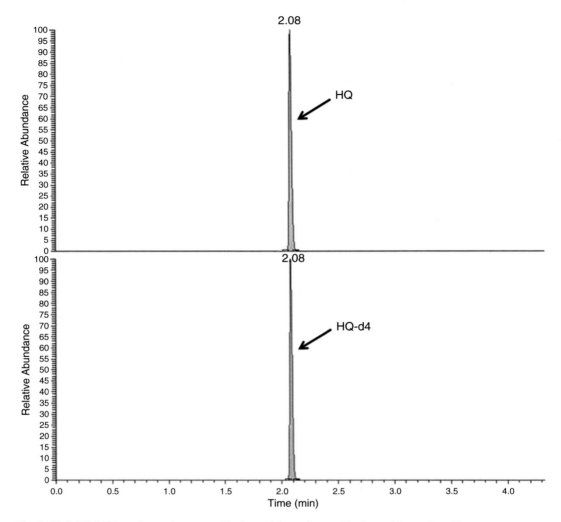

**Fig. 1** TFLC-MS/MS ion chromatograms of hydroxychloroquine and hydroxychloroquine-d4

3. The linear range for the assay is the same as the calibration limits of 15.7–4000 ng/mL with precisions of <5 % over the entire range. The lowest concentration that resulted in a CV of 20 % was determined to be 6.0 ng/mL. However, specimens that are lower than the lowest calibrator are reported as less than that value ("<15.7 ng/mL"). *See* **Notes 3** and **4** for further information regarding linearity and accuracy, respectively.

## 4  Notes

1. A challenge in the development of this assay was significant carryover from one sample to the next. A rigorous validation of carryover was conducted as previously described by running

"high" (2000 ng/mL) and "low" (15.7 ng/mL) spiked controls in various sequences [9]. Inclusion of a 2 min wash step after elution of hydroxychloroquine significantly minimized carry-over, defined as the mean of low–low results subtracted from the mean of high–low results, to within <3 standard deviations of the low–low results.

2. It is important to note the shapes of the standard peaks to verify that the drug has eluted in well resolved symmetrical peaks. If this has not occurred, it is indicative of a problem with the run. Additionally, each sample peak should be evaluated for quality. If a peak is excessively jagged, noisy, or deviates from a bell shape, data for this injection must not be used and troubleshooting must be performed.

3. Though a majority of reported HQ concentrations fall between 50 and 1700 ng/mL, at least one reported level exceeded 2000 ng/mL [10]. For any specimens greater than the highest calibrator, a dilution should be made with drug-free whole blood. Add 100 µL of sample to 100 µL of drug-free EDTA blood, mix gently by inverting, and extract as normal. Dilutions of concentrations as high as 4000 ng/mL were validated with this assay and yielded a CV of within 3 % at each level. Therefore, we have added a 4000 ng/mL high-end calibrator to the assay to cover all anticipated result values.

4. Because a reference method for this assay does not exist, this assay was validated with side-by-side comparisons of samples both analyzed in-house and sent to a reference laboratory for analysis by LC-MS/MS as previously described [9]. Deming regression and statistical analysis yielded a Pearson correlation of 0.9974, indicating excellent correlation over the concentration range of 15–2000 ng/mL. Because no proficiency testing for hydroxychloroquine is available at this time, our laboratory participates in a twice yearly sample exchange with another laboratory performing hydroxychloroquine testing to establish acceptability criteria.

# References

1. Ben-Zvi I, Kivity S, Langevitz P, Shoenfeld Y (2012) Hydroxychloroquine: from malaria to autoimmunity. Clin Rev Allergy Immunol 42(2): 145–153. doi:10.1007/s12016-010-8243-x

2. Tett SE, Cutler DJ, Day RO, Brown KF (1989) Bioavailability of hydroxychloroquine tablets in healthy volunteers. Br J Clin Pharmacol 27(6):771–779

3. Costedoat-Chalumeau N, Amoura Z, Hulot JS, Hammoud HA, Aymard G, Cacoub P, Frances C, Wechsler B, du Huong LT, Ghillani P, Musset L, Lechat P, Piette JC (2006) Low

blood concentration of hydroxychloroquine is a marker for and predictor of disease exacerbations in patients with systemic lupus erythematosus. Arthritis Rheum 54(10):3284–3290. doi:10.1002/art.22156

4. Munster T, Gibbs JP, Shen D, Baethge BA, Botstein GR, Caldwell J, Dietz F, Ettlinger R, Golden HE, Lindsley H, McLaughlin GE, Moreland LW, Roberts WN, Rooney TW, Rothschild B, Sack M, Sebba AI, Weisman M, Welch KE, Yocum D, Furst DE (2002) Hydroxychloroquine concentration-response

relationships in patients with rheumatoid arthritis. Arthritis Rheum 46(6):1460–1469. doi:10.1002/art.10307

5. Frances C, Cosnes A, Duhaut P, Zahr N, Soutou B, Ingen-Housz-Oro S, Bessis D, Chevrant-Breton J, Cordel N, Lipsker D, Costedoat-Chalumeau N (2012) Low blood concentration of hydroxychloroquine in patients with refractory cutaneous lupus erythematosus: a French multicenter prospective study. Arch Dermatol 148(4):479–484. doi:10.1001/archdermatol.2011.2558

6. Brown RR, Stroshane RM, Benziger DP (1986) High-performance liquid chromatographic assay for hydroxychloroquine and three of its major metabolites, desethylhydroxychloroquine, desethylchloroquine and bidesethylchloroquine, in human plasma. J Chromatogr 377:454–459

7. Tett SE, Cutler DJ, Brown KF (1985) High-performance liquid chromatographic assay for hydroxychloroquine and metabolites in blood and plasma, using a stationary phase of poly(styrene divinylbenzene) and a mobile phase at pH 11, with fluorimetric detection. J Chromatogr 344:241–248

8. Wang LZ, Ong RY, Chin TM, Thuya WL, Wan SC, Wong AL, Chan SY, Ho PC, Goh BC (2012) Method development and validation for rapid quantification of hydroxychloroquine in human blood using liquid chromatography-tandem mass spectrometry. J Pharm Biomed Anal 61:86–92. doi:10.1016/j.jpba.2011.11.034

9. Fuzery AK, Breaud AR, Emezienna N, Schools S, Clarke WA (2013) A rapid and reliable method for the quantitation of hydroxychloroquine in serum using turbulent flow liquid chromatography-tandem mass spectrometry. Clin Chim Acta 421:79–84. doi:10.1016/j.cca.2013.02.018

10. Tett SE, Cutler DJ, Day RO, Brown KF (1988) A dose-ranging study of the pharmacokinetics of hydroxy-chloroquine following intravenous administration to healthy volunteers. Br J Clin Pharmacol 26(3):303–313

# Chapter 20

# Quantification of Iohexol in Serum by High-Performance Liquid Chromatography-Tandem Mass Spectrometry (LC-MS/MS)

## Faye B. Vicente, Gina Vespa, Alan Miller, and Shannon Haymond

## Abstract

Iohexol is a nonradioactive contrast medium, and its clearance from serum or urine is used to measure glomerular filtration rate (GFR). GFR is the most useful indicator of kidney function and progression of kidney disease. GFR determination using iohexol clearance is increasingly being applied in clinical practice, given its advantages over and correlation with inulin. We describe a high-performance liquid chromatography tandem mass spectrometry (LC-MS/MS) method for iohexol clearance, requiring only 50 μL of serum. The sample preparation involves protein precipitation with LC/MS-grade methanol, containing ioversol as the internal standard. Samples are centrifuged and supernatant is dried under nitrogen gas at room temperature. Samples are reconstituted with mobile phase (ammonium acetate—formic acid—water). Iohexol is separated using an HPLC gradient method on a C-8 analytical column. MS/MS detection is in the multiple-reaction monitoring (MRM) mode and the transitions monitored are $m/z$ 822.0 to $m/z$ 804.0 and $m/z$ 807.0 to $m/z$ 588.0 for iohexol and ioversol, respectively.

**Key words** Iohexol, Glomerular filtration rate, Mass spectrometry, Liquid chromatography, Serum, Quantification

## 1 Introduction

Measured glomerular filtration rate (mGFR) is the best indicator of renal function in children and adolescents. Accurate assessment of GFR is critical for diagnosing acute and chronic kidney disease, providing early intervention to prevent end-stage renal failure, safely prescribing nephrotoxic and renally cleared drugs, and monitoring for adverse side effects from medications. Estimates of GFR are commonly calculated using equations based on creatinine and other parameters (e.g., BUN, cystatin C, race, gender, weight, and height). Although relatively inexpensive and convenient, creatinine-based clearance is limited due to dependence on muscle mass, relative insensitivity to detect small changes in renal function and the assumption that extra renal clearance of creatinine

Uttam Garg (ed.), *Clinical Applications of Mass Spectrometry in Drug Analysis: Methods and Protocols*, Methods in Molecular Biology, vol. 1383, DOI 10.1007/978-1-4939-3252-8_20, © Springer Science+Business Media New York 2016

is small. In children and adolescents, these equations are particularly problematic [1]. The largest study in children that has directly compared estimating equations with iohexol clearance mGFR showed that the best eGFR formula yielded 87.7 % of eGFR within 30 % of the iohexol-based mGFR and 45.6 % within 10 % [2]. Efforts continue to refine and improve estimating equations for use in pediatrics but there are frequently cases where an accurate and clinically useful method for determination of measured GFR is needed to assess pediatric kidney function. Iohexol (Omnipaque™, GE Healthcare) is an iodinated, water-soluble, nonionic monomeric contrast medium, and it is a suitable marker for GFR, as it is not secreted, metabolized, or reabsorbed by the kidney [3].

The "gold standard" for GFR measurement has been by inulin clearance. However, inulin clearance requires timed urine collections, technically difficult assays and inulin is no longer readily available in the USA. Iohexol clearance to determine GFR is a comparable alternative to inulin these approaches yield highly correlated values for GFR [4, 5]. GFR determination using iohexol clearance has been increasingly accepted and applied in clinical practice because it is accurate, readily available, nonradioactive, safe, and used intravenously even in the presence of renal disease [2, 6]. This chapter describes a high-performance liquid chromatography tandem mass spectrometry (LC-MS/MS) method for iohexol clearance, requiring only 50 µL of serum.

## 2  Materials

### 2.1  Samples

Serum with no gel separator. Samples are stable for 24 h at 4 °C and 3–4 months at –20 °C up to three freeze/thaw cycles.

### 2.2  Solvents and Reagents

1. Special reagent water (SRW) obtained from Millipore Milli-Q Integral 5 Water Purification System.

2. Ammonium acetate HPLC grade, 1 M, prepared with special reagent water. Stable at 4 °C for 1 month.

3. Mobile phase A and purge solvent (2 mM ammonium acetate/0.1 % (v/v) formic acid in SRW): Add 2 mL of 1 M ammonium acetate solution and 1 mL formic acid to 1 L water. Stable at room temperature for 2 weeks.

4. Mobile phase B (2 mM ammonium acetate/0.1 % (v/v) formic acid in methanol): Add 2 mL of 1 M ammonium acetate solution and add 1 mL formic acid to 1 L methanol. Stable at room temperature for 2 weeks.

5. Column wash solvent (50 % methanol in water): Mix 500 mL of water and 500 mL of methanol in a 1-L solution bottle. Stable at room temperature for 1 month.

6. Needle wash solvent (100 % methanol): Stable at room temperature for 1 month.

7. Charcoal dextran-stripped human serum.

**2.3 Standards and Calibrators**

1. Primary standard: Iohexol (U.S. Pharmacopeial Convention).

2. Iohexol calibrator stock solutions (100–20,000 µg/mL primary standard in SRW):

   (a) Add 200 mg primary standard to 10 mL volumetric flask, bring to volume with SRW, and mix well by inversion. This iohexol standard stock solution 1 is 20,000 µg/mL. Stable at −70 °C for 2 years.

   (b) Add 5 mL of the previous solution to 10 mL volumetric flask, bring to volume with SRW, and mix well by inversion. This iohexol standard stock solution 2 is 10,000 µg/mL. Stable at −70 °C for 2 years.

   (c) Add 0.5 mL of standard stock solution 1 to 10 mL volumetric flask, bring to volume with SRW, and mix well by inversion. This iohexol standard stock solution 3 is 1000 µg/mL. Stable at −70 °C for 2 years.

   (d) Add 1 mL of standard stock solution 3 to 10 mL volumetric flask, bring to volume with SRW, and mix well by inversion. This iohexol standard stock solution 4 is 100 µg/mL. Stable at −70 °C for 2 years.

3. Calibrators (10–1000 µg/mL in SRW): Prepare calibrators 1–4 by diluting the standard stock solutions according to Table 1. For each dilution step: Add appropriate amount of standard stock solution(s) as shown in Table 1 to 10 mL volumetric flask(s) and bring to volume with serum. Mix well by inversion after each dilution step. Stable at −70 °C for 2 years (*see* **Note 1**).

4. HPLC/MS check standard (0.2 µg/mL in SRW): Add 0.5 mL of standard stock solution 4 to 250 mL volumetric flask, bring to volume with SRW, and mix well by inversion. Stable at −70 °C for 2 years.

**Table 1**
**Preparation of calibrators**

| Calibrator | Volume of standard stock solution (mL) | Volume of special reagent water (mL) | Final concentration (µg/mL) |
|---|---|---|---|
| 1 | 0.5 (stock solution 1) | 9.5 | 1000 |
| 2 | 0.5 (stock solution 2) | 9.5 | 500 |
| 3 | 1.0 (stock solution 3) | 9.0 | 100 |
| 4 | 1.0 (stock solution 4) | 9.0 | 10 |

**2.4  Internal Standard and Quality Controls**

1. Primary internal standard (I.S.): Ioversol (U.S. Pharmacopeial Convention).

2. I.S. working solution/protein precipitation solution (40 µg/mL primary I.S. in methanol):

    (a) Add 50 mg of primary I.S. to 10 mL volumetric flask; bring to volume with methanol and mix. This intermediate solution is 5000 µg/mL. Stable at −70 °C for 2 years.

    (b) Add 2 mL of previous solution to 250 mL volumetric flask, bring to volume with methanol and mix. Stable at −70 °C for 2 years or −20 °C for 3 months.

3. Iohexol quality control stock solutions (650–6500 µg/mL primary standard in serum):

    (a) Primary standard is separately weighed or from a different lot than that used for calibrators.

    (b) Add 65 mg of primary standard to 10-mL volumetric flask, bring to volume with special reagent water, and mix well by inversion. This iohexol quality control stock solution 1 is 6500 µg/mL. Stable at −70 °C for 2 years.

    (c) Add 1 mL of previous solution to 10 mL volumetric flask, bring to volume with special reagent water and mix well by inversion. This iohexol quality control stock solution is 650 µg/mL. Stable at −70 °C for 2 years.

4. Quality controls (13, 130, 780 µg/mL in serum): Prepare low, medium, and high controls by diluting quality control stock solutions as shown in Table 2. For each dilution step: Add appropriate amount of quality control stock solutions as shown in Table 2 to 25 mL volumetric flask and bring to volume with serum. Mix well by inversion after each dilution step. Stable at −70 °C for 2 years (*see* **Note 1**).

**2.5  Supplies**

1. 12 × 32 mm glass screw cap with conical bottom HPLC vials.

2. 1.2 mL polypropylene cryogenic vials. These are used to store calibrators, controls, and check standard.

**Table 2**
**Preparation of quality controls**

| Quality control | Volume of control stock solution (mL) | Drug-free serum (mL) | Final concentration (µg/mL) |
|---|---|---|---|
| Low | 0.5 (stock solution 2) | 24.5 | 13 |
| Medium | 0.5 (stock solution 1) | 24.5 | 130 |
| High | 3.0 (stock solution 1) | 24.5 | 780 |

3. Analytical column: Phenomenex Luna C8, 3 μm 50 × 3.0 mm I.D.

4. Guard column: Phenomenex C8, 4 × 2.0 mm I.D.

**2.6 Equipment**

1. Waters 2795 Alliance HT Separation Module with Micromass Quattro Micro API equipped with MassLynx.

2. Hamilton MicroLab Nimbus4 automated multi-channel pipetting workstation.

3. Thermo Scientific Reacti-Therm III Heating/Stirring Module.

# 3 Methods

**3.1 Stepwise Procedure**

1. Ensure that the instrument is properly tuned and verify system performance (see **Notes 2** and **3**).

2. Pipette 50 μL of sample (calibrators, quality controls, serum blank, and patient sera) to labeled 1.5 mL microcentrifuge tubes (see **Notes 4** and **5**).

3. Add 400 μL of the precipitation solution (see **Note 4**).

4. Cap and vortex mix tubes for 30 s.

5. Centrifuge at 9015 rcf or higher for 1 min at room temperature.

6. Transfer 100 μL supernatant into labeled 13 × 100 mm glass culture tubes (see **Note 4**).

7. Using the Thermo Scientific Reacti-Therm III Heating/ Stirring Module, dry samples under nitrogen gas at room temperature for 15 min (or until completely dry) (see **Note 6**).

8. Reconstitute the supernatant by adding 2.0 mL mobile phase A (see **Note 4**).

9. Cap the tubes and vortex mix thoroughly 10 s (see **Note 4**).

10. Centrifuge at 1430–1500 rcf for 5 min.

11. Transfer solution to appropriately labeled autosampler vials (see **Note 4**).

12. Inject 3 μL of sample onto LC-MS/MS.

**3.2 Analysis**

1. Instrumental operating parameters are given in Table 3 (see **Notes 7–9**).

2. Analyze the data using the QuanLynx software (Waters Corporation).

3. With each analytical run, a 4-point standard calibration curve is created by linear regression forced to the origin based on iohexol/internal standard peak area ratios using the quantifying

**Table 3**
**LC-MS/MS operating conditions**

| A. HPLCa | | | |
|---|---|---|---|
| Column temperature | Room temperature | | |
| Flow rate | 0.500 mL/min | | |
| Gradient | Time (min) | Mobile phase A (%) | Curve |
| | 0.00 | 98.0 | 1 |
| | 3.00 | 0.0 | 6 |
| | 5.00 | 0.0 | 6 |
| | 5.50 | 98.0 | 6 |
| **B. MS/MS tune settingsb** | | | |
| Capillary voltage (kV) | 1.4 | | |
| Source temperature (°C) | 130 | | |
| Desolvation temperature (°C) | 400 | | |
| Cone gas (L/h) | 35 | | |
| Desolvation gas (L/h) | 700 | | |
| Collision gas pressure (mbar) | 3.70 e-3 | | |
| LM1 resolution | 14.5 | | |
| HM1 resolution | 14.5 | | |
| Ion energy 1 | 0.5 | | |
| MS/MS entrance | −2 | | |
| MS/MS exit | 1 | | |
| LM2 resolution | 13.2 | | |
| HM2 resolution | 13.2 | | |
| Ion energy 2 | 2.3 | | |

[a]The total run time is 7.5 min. Solvent flow was diverted from the source to waste at 0–1 min and at 5–7.5 min
[b]Tune settings may vary slightly between instruments

ions indicated in Table 4. The concentrations of the controls and unknown samples are determined from the curve.

4. The expected retention times for iohexol and ioversol are 2.36 min (acceptable range: 2.16–2.56 min) and 2.14 min (acceptable range: 1.94–2.34 min), respectively. Representative ion chromatograms for iohexol and I.S. are shown in Fig. 1.

5. Verify the performance during the analytical run by monitoring the internal standard peak area. An acceptable limit should be defined during method development or validation.

**Table 4**
**MRM method parameters for iohexol and ioversol**

| Analyte | Precursor ion (M+H)⁺ | Product ion | Dwell (s) | Cone (V) | Collision (eV) | Interchannel delay (s) | Inter-scan delay (s) |
|---------|---------------------|-------------|-----------|----------|----------------|------------------------|----------------------|
| Iohexol | 821.9 | 803.7 | 0.2 | 38 | 21 | 0.03 | 0.03 |
| Ioversol | 807.9 | 588.7 | 0.2 | 38 | 24 | 0.03 | 0.03 |

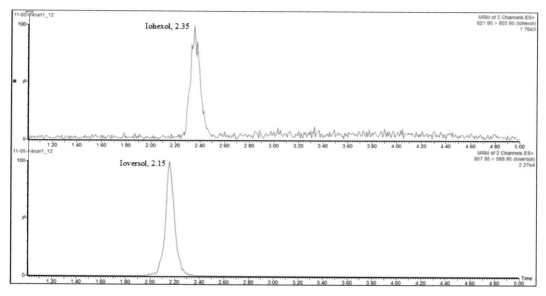

**Fig. 1** Representative LC-MS/MS ion chromatograms of iohexol (21.5 µg/mL) and ioversol (39.7 µg/mL) in human serum

We determined 850 to be the minimum acceptable IS peak area in our method. Re-inject the sample if the internal standard peak area is below the acceptance limit. If after re-injection, the internal standard peak area is still below the limit, determine the signal-to-noise ratio of the iohexol peak. Signal-to-noise ratio greater than 10 is acceptable for reporting.

6. Evaluate for carryover effects in the serum blank injected after Calibrator 1. Carryover is significant when iohexol concentration in the serum blank is greater than 3.2 µg/mL and in the low-quality control injected immediately after the high-quality control is greater than the two standard deviations of the target value and/or assigned mean. 3.2 µg/mL was selected as the carryover limit since it was the lowest detectable concentration determined during method validation. If carryover is significant, troubleshoot and perform corrective action. Repeat the evaluation to demonstrate that carryover is no longer detected.

7. Run is acceptable if the calculated concentrations in the control samples are within two standard deviations of the target values and/or assigned means.

8. The method is linear from 7.8 to 2000.0 μg/mL with intra- and inter-day precision of <4 %. Correlation with a reference method (Mario Negri Institute, Italy) demonstrated good agreement with a slope of 1.019, an intercept of −1.385 and a correlation coefficient of 0.996.

9. No significant ion suppression or matrix effect was found in charcoal-stripped serum (used for preparing calibrators and quality controls), hemolyzed and lipemic serum, and pooled serum from kidney diseases and general pediatric patients.

## 4  Notes

1. Calibrators and quality controls are pre-aliquoted and stored in −70 °C until use. Pipette 125 μL of the calibrator and quality controls solutions into 1.2 mL cryogenic vials. Opened vials are stable for 14 days at −20 °C.

2. Tuning the mass spectrometer: To adjust the mass spectrometer parameters for optimum sensitivity and stability of ions measured, 10 μg/mL tuning solutions of iohexol and ioversol are infused into the ion source at 10 μL/min while solvent from the HPLC consisting of 20 % Mobile Phase A and 80 % Mobile Phase B is introduced via a peak "tee" connector at 0.500 mL/min.

3. System check: To verify system performance before running patient samples, inject the Iohexol HPLC/MS check standard solution after a water blank. Verify that the iohexol retention time is within acceptable limits and that the signal-to-noise (peak-to-peak) of the iohexol peak is acceptable. The HPLC/MS check standard solution is pre-aliquoted and stored in −70 °C until use. Opened vials are for one time use only.

4. **Steps 1, 2, 5, 7, 8**, and **10** in Subheading 3.1 can be performed using an automated liquid handling system.

5. When using the automated liquid handling system for pipetting tasks, verify by visual inspection that all samples have been dispensed accurately before proceeding to the next step. Do not vigorously mix or vortex the serum sample as this can cause bubble formation at the surface of the specimen which may cause inaccurate sampling and measurement. Care must be taken to remove these bubbles before analysis begins. This can be done by poking the bubbles with a wooden stick, or by a short (5 min) centrifugation at $1500 \times g$. When pouring protein precipitation or reconstituting solvent into the reservoirs,

ensure that there are no bubbles on the surface of the liquid to prevent pipetting problems.

6. Verify that the samples have been completely dried down by visually inspecting the bottom of the tubes.

7. After analytical runs are completed, the column is flushed for 45 min at a flow rate of 0.250 mL/min and stored with 70 % methanol in water.

8. The product ions selected for the multiple reaction monitoring (MRM) experiment were the most stable and abundant peaks observed during MS optimization, corresponding to a water loss (m/z 18) for iohexol and a loss of m/z 219 for ioversol. Although the chosen transitions are different, validation data indicate that this does not impact the quantitation of iohexol.

9. The use of ioversol as an internal standard has limitations in that it is not a stable isotope-labeled form of iohexol; therefore, its physicochemical properties may not be completely consistent with that of iohexol.

## References

1. Zappitelli M, Parvex P, Joseph L, Paradis G, Grey V, Lau S, Bell L (2006) Derivation and validation of cystatin C-based prediction equations for GFR in children. Am J Kidney Dis 48:221–230

2. Schwartz GJ, Furth S, Cole SR, Warady B, Munoz A (2006) Glomerular filtration rate via plasma iohexol disappearance: pilot study for chronic kidney disease in children. Kidney Int 69:2070–2077

3. Olsson B, Aulie A, Sveen K, Andrew E (1983) Human pharmacokinetics of iohexol. A new non-ionic contrast medium. Invest Radiol 18:177–182

4. Brown SC, O'Reilly PH (1991) Iohexol clearance for the determination of glomerular filtration rate in clinical practice: evidence for a new gold standard. J Urol 146:675–679

5. Gaspari F, Perico N, Ruggenenti P, Mosconi L, Amuchastegui CS, Guerini E, Daina E, Remuzzi G (1995) Plasma clearance of nonradioactive iohexol as a measure of glomerular filtration rate. J Am Soc Nephrol 6:257–263

6. Schwartz GJ, Work DF (2009) Measurement and estimation of GFR in children and adolescents. Clin J Am Soc Nephrol 4:1832–1843

# Chapter 21

# Quantitation of Teriflunomide in Human Serum/Plasma Across a 40,000-Fold Concentration Range by LC/MS/MS

## Geoffrey S. Rule, Alan L. Rockwood, and Kamisha L. Johnson-Davis

## Abstract

Leflunomide is a prodrug used primarily for treatment of rheumatoid arthritis. The active metabolite, teriflunomide (A77 1726), inhibits the enzyme dihydroorotate dehydrogenase and thereby reduces the synthesis of pyrimidine ribonucleotides. Teriflunomide is also administered directly and finds use in treating multiple sclerosis. Therapeutic concentrations are generally in the tens of µg/mL serum or plasma and, due to adverse effects and the time required to reach steady state, therapeutic drug monitoring is beneficial. The drug is also a potential teratogen.

A method was developed and validated to quantify the drug teriflunomide over a 40,000-fold concentration range of 5 ng/mL to 200 µg/mL in serum or plasma. This is accomplished by dividing the quantitative range into two separate but overlapping regions; a high curve and a low curve range. Samples are evaluated first against the high curve after a 100-fold dilution of the sample extract. Samples falling below the upper curve region are evaluated again without dilution and quantified, if possible, against the low curve calibration standards. Appropriate choice of a concentration for the deuterated internal standard (D4-teriflunomide) allows for a single, identical, extraction procedure to be performed for both curve regions but with the dilution performed for high curve samples. The method is rugged and reliable with good accuracy and precision statistics.

**Key words** Leflunomide, Teriflunomide, Dynamic range, LC/MS/MS, Arthritis

## 1 Introduction

Leflunomide is a prodrug approved by the FDA in 1998 and brought to market by Sanofi-Aventis under the name Arava®. It is administered largely for the treatment of rheumatoid arthritis and is classed as one of the disease-modifying antirheumatic drugs, or DMARDs. It has been shown not only to improve the quality of life in patients, by virtue of allowing better physical activity, but also to slow disease progression [1]. Leflunomide contains an isoxazole ring which is opened nonenzymatically to the active metabolite, teriflunomide. Another drug formulation, marketed as Aubagio®, consists of the ring-opened form and is used treatment of active relapsing–remitting multiple sclerosis [2]. Leflunomide

Uttam Garg (ed.), *Clinical Applications of Mass Spectrometry in Drug Analysis: Methods and Protocols*, Methods in Molecular Biology, vol. 1383, DOI 10.1007/978-1-4939-3252-8_21, © Springer Science+Business Media New York 2016

has also been prescribed, off-label, for other uses in renal transplant patients, treating cytomegalovirus viremia [3], and BK virus-associated nephropathy [4], and for treatment in spondylo and psoriatic arthritis [5].

Due to metabolism, circulating levels of leflunomide after oral administration are generally very low in comparison with levels of teriflunomide [6]. Both leflunomide and teriflunomide have the same molecular weight and, after collision induced dissociation of the protonated parent by tandem mass spectrometry, the same two most abundant product ions. Consequently, it was thought desirable to obtain a chromatographic separation of leflunomide from teriflunomide to eliminate the possibility of interference in determination of circulating levels of the two drug forms.

In addition, teriflunomide is thought to have some potential for teratogenic activity based on animal studies. Therapeutic doses of the drug generally yield plasma or serum concentrations in the tens of micrograms per milliliter. Because of the possible teratogenic activity however, it is recommended that those wishing to conceive offspring achieve levels lower than 20 ng/mL prior to conception as determined on two separate occasions [7]. To complicate the clearance of the drug, it has a fairly lengthy half-life of approximately 2 weeks due largely to enterohepatic circulation. In some cases levels of the metabolite have been found in individuals up to a year after ceasing therapy. To assist clearance several options are available including use of activated charcoal and cholestyramine.

Due to the wide range of possible concentration, from several hundred µg/mL down to the medical decision point of 20 ng/mL, it was important for us to develop a method that would cover a very broad concentration range. Here we describe a method that covers a 40,000-fold range from 5 ng/mL to 200 µg/mL through use of two separate but overlapping calibration curves. A single, identical extraction procedure is utilized for both curve ranges, with the exception that samples analyzed for the higher concentration range are diluted 100-fold prior to analysis. Earlier methods for determination of teriflunomide or leflunomide have been described in the literature as using HPLC/UV [6, 8, 9], LC/MS/MS [10].

## 2  Materials

### 2.1  Samples

Heparinized or EDTA plasma, plain (red top) serum: Collect and process specimens according to standard phlebotomy procedures. Freeze specimens at –20 °C or colder and transport.

### 2.2  Reagents

1. Dimethylsulfoxide (DMSO): Sigma-Aldrich (St. Louis, MO).

2. Mobile phase A: 0.1 % formic acid in deionized water.

3. Mobile phase B: 0.1 % formic acid in water:methanol:acetonitrile (0.5:0.5:9).

4. Control serum: Lampire Biological Laboratories (Pipersville, PA).

5. Double-blank solution: 0.1 % formic acid in methanol:acetonitrile (1:1).

6. Dilution solution: 0.1 % formic acid in 25 % water, 75 % methanol:acetonitrile (1:1).

7. Autosampler wash solvents:

    (a) Wash 1, methanol:water (4:1) containing 0.1 % trifluoro-acetic acid.

    (b) Wash 2, methanol:water (2:3).

**2.3 Standards and Internal Standards**

1 Primary standards: Leflunomide and teriflunomide (Toronto Research Chemicals, North York, ON).

2 Primary internal standard: D4-teriflunomide (Toronto Research Chemicals, North York, ON).

3 Teriflunomide standard stock solution: 10 mg/mL in DMSO.

4 Working internal standard/ protein precipitating solution: 333 ng/mL D4-teriflunomide in methanol:acetonitrile (1:1) containing 0.1 % formic acid.

**2.4 Calibrators and Controls**

1. Calibrators: Prepare high curve calibrators (Std 6–10), starting with preparation of Std 10 from stock solution, according to the scheme shown in Table 1. All remaining calibrators are prepared from Std 10, or dilutions thereof. Prepare low curve calibrators (Std 1–5), similarly, according to Table 1.

2. Controls: Prepare controls D, E, and F according to Table 2 from a separate preparation of stock solution. Prepare controls A, B, and C according to Table 2 by dilution of controls D and E. Vortex mix and aliquot into microcentrifuge tubes. Store at −20 C or lower until use.

**2.5 Analytical Equipment and Supplies**

1. Agilent 1200 series pump, CTC autosampler, and AB Sciex API4000 triple-quadrupole mass spectrometer with Analyst software (version 1.5.1).

2. Analytical column; 2 mm × 10 cm, Luna PFP [2] (pentafluoro-phenyl) phase on 3 μm particles, Phenomenex (Torrance, CA).

3. 1.7 mL microcentrifuge tubes.

4. Centrifuge to accommodate microcentrifuge tubes and capable of achieving $10,000 \times g$.

5. Autosampler vials.

**Table 1**
**Standard curve preparation for high and low curve regions**

| High curve | | | | |
|---|---|---|---|---|
| | Concentration (µg/mL) | Volume (µL) | Stock or std | Control serum (µL) |
| Std 10 | 200 | 5 | Stock | 245 |
| Std 9 | 50 | 40 | Std 10 | 120 |
| Std 8 | 10 | 10 | Std 10 | 190 |
| Std 7 | 2 | 10 | Std 10 | 990 |
| Std 6 | 0.8 | 100 | Std 7 | 150 |
| **Low curve** | | | | |
| | Concentration (µg/mL) | Volume (µL) | Std | Control serum (µL) |
| Std 5 | 1 | 200 | Std 7 | 200 |
| Std 4 | 0.5 | 100 | Std 5 | 100 |
| Std 3 | 0.1 | 25 | Std 4 | 100 |
| Std 2 | 0.02 | 10 | Std 4 | 240 |
| Std 1 | 0.005 | 5 | Std 4 | 495 |

**Table 2**
**Quality control sample preparation for validation**

| Low curve | | | High curve | | |
|---|---|---|---|---|---|
| | | Conc. (µg/mL) | | | Conc. (µg/mL) |
| | Low control = A | 0.02 | | Low control = D | 1 |
| | Medium control = B | 0.1 | | Medium control = E | 80 |
| | High control = C | 0.8 | | High control = F | 170 |
| Control | Vol. flask (mL) | | Control | Vol. flask (mL) | |
| A | 10 | Add 0.2 µg as 200 µL Control D | D | 10 | Add 10 ug as 125 µL Control E |
| B | 10 | Add 1 µg as 12.5 µL Control E | E | 10 | Add 800 ug as 80 µL 10 mg/mL Stock |
| C | 10 | Add 8 µg as 100 µL of Control E | F | 10 | Add 1.7 mg as 170 µL 10 mg/mL Stock |

# 3 Methods

### 3.1 Stepwise Procedure

1. Add 100 μL blank serum, standard, control, or patient serum/ plasma to labeled microcentrifuge tubes. (*Note:* For each curve range both a *blank* and *double-blank* sample are prepared.) A *blank* is analyzed as the first and last injection of each sequence and consists of a known negative serum sample. A *double blank* is analyzed after each high standard. A *double blank* sample is one prepared in identical fashion to the ordinary *blank* but in the absence of internal standard. Placing this sample after the high calibration standard allows one to distinguish between autosampler carryover and unlabeled analyte that may be contributed through the internal standard addition as an impurity (*see* **Note 1**).

2. Add 300 μL of "double-blank solution" to the *double blanks.* Add 300 μL of "working internal standard" to all other calibrators, controls, and patient samples (*see* **Note 2**).

3. Cap tubes and vortex mix for 3 min.

4. Centrifuge for 10 min at $10,000 \times g$.

5. For high curve: Transfer 2 μL of the supernatant into autosampler vials containing 200 μL of "dilution solution," and cap.

6. For low curve: Transfer 100 μL of supernatant into autosampler vials and cap.

7. Transfer vials to autosampler. Inject 3 μL of each sample in sequence.

### 3.2 Analysis

1. HPLC conditions, gradient, and MS parameters are as shown in Tables 3 and 4 (*see* **Notes 3–5**).

2. Data were collected and analyzed using AB Sciex Analyst® software (version 1.5.2). Example extracted ion current profiles for the low curve Std 1 and internal standard are shown in Fig. 1.

3. Linear regression analysis of the calibration standards was made by $1/x^2$ weighted regression of peak area ratio versus analyte concentration.

# 4 Notes

1. Prepare both sets of calibration standards, all quality control samples, and patient samples in the same batch. Prepare each patient sample with 100× dilution, as described, and evaluate against the high curve calibration standards. If within the high calibration curve region, report the determined value. (If sample is above the ULOQ, extract sample a second time with

**Table 3**
**LC/MS/MS operating parameters**

| A.   HPLC | | | | |
|---|---|---|---|---|
| Column temp. 40 ° C | | | | |
| *Step* | *Total time* (min) | *Flow rate* (µl/min) | *A* (%) | *B* (%) |
| 0 | 0.0 | 300 | 30 | 70 |
| 1 | 0.1 | 300 | 30 | 70 |
| 2 | 0.6 | 300 | 0 | 100 |
| 3 | 3.0 | 300 | 0 | 100 |
| 4 | 3.1 | 300 | 30 | 70 |
| 5 | 3.2 | 300 | 30 | 70 |
| B.   *MS/MS parameter table* | | | | |
| CUR: 25 | | | | |
| TEM:500 | | | | |
| GS1: 35 | | | | |
| GS2: 30 | | | | |
| Nitrogen gas is used for both GS1 and GS2 | | | | |
| ihe: ON | | | | |
| IS: –4300 V | | | | |
| CAD: Medium | | | | |
| DP: –55 | | | | |
| EP: –10 | | | | |
| Detector parameters (negative): CEM 2500 | | | | |

**Table 4**
**List of precursor and product ions for teriflunomide and internal standard**

| | Precursor ion | Product ion | Dwell (ms) | Collision energy (eV) |
|---|---|---|---|---|
| Teriflunomide | 269.1 | 82.0 | 50 | 27 |
| | 269.1 | 160.1 | 50 | 34 |
| D4-teriflunomide | 273.1 | 82.0 | 50 | 27 |
| | 273.1 | 164.1 | 50 | 34 |

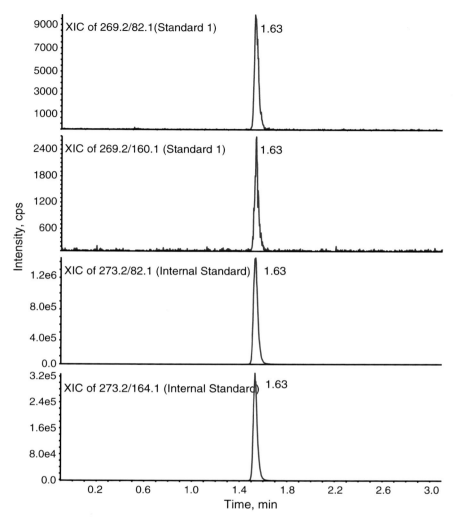

**Fig. 1** Extracted ion current chromatograms of teriflunomide in Standard 1 (5 ng/mL), and internal standard (D4-teriflunomide), showing two transitions for each

appropriate matrix dilution.) If the patient sample is below the LLOQ of the upper curve range then prepare and analyze the undiluted sample extract along with the low curve calibrators and quality control samples. Report the resulting value as appropriate.

2. The concentration of internal standard utilized is equivalent to 1 μg/mL in the plasma/serum sample. This concentration is near the low end of the upper curve and at the high end of the lower curve. It is possible to use this same solution (concentration) for both curves since there is a lack of a significant analyte isotope interference with the IS and because only a small amount of unlabeled analyte is present in the IS. We discuss this topic elsewhere in detail [11].

3. The mass spectrometer is infused with a solution of naproxen (1 ng/μL in 1:1 methanol:acetonitrile, at 10 μL/min) prior to each run to verify mass spectrometer performance in terms of sensitivity and spectral resolution. The parent ion at m/z 229.1 is evaluated by single MS (Q1) and the fragment ion at m/z 185.1 evaluated by MS/MS (for Q3). The instrument is operated at peak widths of 0.7 amu +/−0.1 amu, at half height, and mass accuracy of +/−0.1 amu.

4. A test injection is made with a solution of teriflunomide and leflunomide (each at 100 pg/ μL in 1:3, water:acetonitrile) to verify instrument performance (sensitivity and retention time) and chromatographic separation (resolution) prior to each batch of samples.

5. Although the parent drug, leflunomide, is generally found only at very low concentrations, if at all, it is separated chromatographically from the metabolite, teriflunomide, due to the fact that the two have both the same precursor and product ion masses.

## Acknowledgements

The authors would like to thank the ARUP Institute for Clinical and Experimental Pathology for making this work possible.

## References

1. Schattenkirchner M (2000) The use of leflunomide in the treatment of rheumatoid arthritis: an experimental and clinical review. Immunopharmacology 47(2–3):291–298

2. Oh J, O'Connor PW (2013) Teriflunomide for the treatment of multiple sclerosis. Semin Neurol 33(1):45–55

3. Chacko B, John GT (2012) Leflunomide for cytomegalovirus: bench to bedside. Transpl Infect Dis 14(2):111–120

4. Teschner S, Gerke P, Geyer M, Wilpert J, Krumme B, Benzing T et al (2009) Leflunomide therapy for polyomavirus-induced allograft nephropathy: efficient BK virus elimination without increased risk of rejection. Transplant Proc 41(6):2533–2538

5. Babic-Naglic D, Anic B, Novak S, Grazio S, Martinavic Kaliterna D (2010) Treatment of rheumatoid and psoriatic arthritis—review of leflunomide. Reumatizam 57(2):161–162

6. Schmidt A, Schwind B, Gillich M, Brune K, Hinz B (2003) Simultaneous determination of leflunomide and its active metabolite, A77 1726, in human plasma by high-performance liquid chromatography. Biomed Chromatogr 17(4):276–281

7. Brent RL (2001) Teratogen update: reproductive risks of leflunomide (Arava); a pyrimidine synthesis inhibitor: counseling women taking leflunomide before or during pregnancy and men taking leflunomide who are contemplating fathering a child. Teratology 63(2):106–112

8. Chan V, Charles BG, Tett SE (2004) Rapid determination of the active leflunomide metabolite A77 1726 in human plasma by high-performance liquid chromatography. J Chromatogr B Analyt Technol Biomed Life Sci 803(2):331–335

9. van Roon EN, Yska JP, Raemaekers J, Jansen TL, van Wanrooy M, Brouwers JR (2004) A rapid and simple determination of A77 1726 in human serum by high-performance liquid chromatography and its application for optimization of leflunomide therapy. J Pharm Biomed Anal 36(1):17–22

10. Parekh JM, Vaghela RN, Sutariya DK, Sanyal M, Yadav M, Shrivastav PS (2010)

Chromatographic separation and sensitive determination of teriflunomide, an active metabolite of leflunomide in human plasma by liquid chromatography-tandem mass spectrometry. J Chromatogr B Analyt Technol Biomed Life Sci 878(24):2217–2225

11. Rule GS, Clark ZD, Yue B, Rockwood AL (2013) Correction for isotopic interferences between analyte and internal standard in quantitative mass spectrometry by a nonlinear calibration function. Anal Chem 85(8):3879–3885

# Chapter 22

# Determination of Menthol in Plasma and Urine by Gas Chromatography/Mass Spectrometry (GC/MS)

Judy Peat, Clint Frazee, Gregory Kearns, and Uttam Garg

## Abstract

Menthol, a monoterpene, is a principal component of peppermint oil and is used extensively in consumer products as a flavoring aid. It is also commonly used medicinally as a topical skin coolant; to treat inflammation of the mucous membranes, digestive problems, and irritable bowel syndrome (IBS); and in preventing spasms during endoscopy and for its spasmolytic effect on the smooth muscle of the gastrointestinal tract. Menthol has a half life of 3–6 h and is rapidly metabolized to menthol glucuronide which is detectable in urine and serum following menthol use. We describe a method for the determination of total menthol in human plasma and urine using liquid/liquid extraction, gas chromatography/mass spectrometry (GC/MS) in selected ion monitoring mode and menthol-d4 as the internal standard. Controls are prepared with menthol glucuronide and all samples undergo enzymatic hydrolysis for the quantification of total menthol. The method has a linear range of 5–1000 ng/mL, and coefficient of variation <10 %.

**Key words** Menthol, GCMS, Peppermint oil, Irritable bowel syndrome

## 1 Introduction

Menthol is a monoterpene derived from peppermint oil, but can also be prepared synthetically. It is widely known for its use as a topical and oral anesthetic [1]. Also due to its spasmolytic effect in the gastrointestinal tract, it is used in the treatment of irritable bowel syndrome (IBS) [2–6]. Despite the increasingly widespread use of peppermint oil and the treatment of patients with IBS, there are currently no pharmacokinetic or pharmacodynamic data of the principal component, menthol, in pediatric patients. Therefore, there is a current need for menthol assays to support pharmacokinetics and pharmacodynamics studies [7]. Determination of menthol concentration in different pharmaceutical formulations or in peppermint oil may also be indicated [8]. Various methods including gas chromatography with flame ionization or mass spectrometry and high-performance liquid chromatography have been described in the literature [9–14]. Gas chromatography with a flame ionization

Uttam Garg (ed.), *Clinical Applications of Mass Spectrometry in Drug Analysis: Methods and Protocols*, Methods in Molecular Biology, vol. 1383, DOI 10.1007/978-1-4939-3252-8_22, © Springer Science+Business Media New York 2016

detector has historically been the most widely used method for determining menthol levels [10–12]. More recently gas chromatography mass spectrometry methods that are laborious and time consuming have been described in the literature [13, 14]. We describe a novel gas chromatography mass spectrometry method which utilizes a simple liquid/liquid extraction and deuterated internal standard for the quantitation of total menthol in human urine and plasma.

## 2    Materials

### 2.1    Samples

Heparinized plasma, or urine. Store in freezer (<–20 °C) until analysis.

### 2.2    Reagents

1. 0.4 M Sodium phosphate buffer (pH 6.0):

   (a) Phosphate buffer A: Dissolve 11.0 g monobasic sodium phosphate (ACS certified) into 200 mL deionized water.

   (b) Phosphate buffer B: Dissolve 10.7 g dibasic sodium phosphate buffer (ACS certified) into 100 mL deionized water.

   (c) Add phosphate buffer A to a 500 mL beaker. Adjust the pH to 6.0 +/–0.1 by slowly adding phosphate buffer B (see **Note 1**). Stable for 1 year at room temperature.

2. 3.3 M sodium acetate buffer (pH 4.8): Add 2.72 g sodium acetate (ACS certified) to a 10 mL volumetric flask. Add 8 mL deionized water. Adjust pH to 4.8 with acetic acid and qs to 10 mL. Stable 6 months at room temperature.

3. β-Glucuronidase (Sigma, Type H-3: from Helix pomatia, store at 2–8 °C).

4. Human drug-free pooled normal plasma.

### 2.3    Standards and Internal Standards

1. Menthol powder (Toronto Research Chemical, Canada).

2. Menthol β-D-glucuronide powder (Toronto Research Chemical, Canada).

3. Menthol-d4 powder (Toronto Research Chemical, Canada).

4. 1 mg/mL menthol standard stock solution: Add 10 mg of menthol to a 10 mL volumetric flask, dissolve, and bring to volume with methanol.

5. Menthol working standard solutions: Make serial 1:10 dilutions of stock solution with methanol for 100 μg/mL, 10 μg/mL, and 1 μg/mL standards.

6. 100 μg/mL menthol glucuronide stock standard: Add 1 mg of menthol glucuronide to 10 mL volumetric flask and qs with methanol.

7. 10 µg/mL menthol glucuronide working standard: Make a 1:10 dilution of stock.

8. 100 µg/mL internal standard (IS) stock solution: Add 1 mg of menthol-d4 to a 10 mL volumetric flask and qs with methanol.

9. 10 µg/mL IS working solution: Make a 1:10 dilution of stock IS solution with methanol.

Store all stock and working solutions at <−20 °C. Stable for 3 months.

**2.4 Calibrators and Controls**

1. Calibrators: Prepare calibrators 1–7 according to Table 1.

2. Quality controls: Prepare Control 1–3 according to Table 2.

For calibrator and controls add appropriate amount of standard(s) to 10 mL volumetric flask and qs to 10 mL with drug-free plasma.

**Table 1**
**Preparation of calibrators in drug-free plasma**

| Calibrator (ng/mL) | Final volume (mL) | µL of 1 µg/mL menthol standard | µL of 10 µg/mL menthol standard | µL of 100 µg/mL menthol standard |
|---|---|---|---|---|
| Blank | 10 | | | |
| 5 | 10 | 50 | | |
| 10 | 10 | 100 | | |
| 25 | 10 | | 25 | |
| 50 | 10 | | 50 | |
| 100 | 10 | | 100 | |
| 500 | 10 | | | 50 |
| 1000 | 10 | | | 100 |

**Table 2**
**Preparation of quality controls made in drug-free plasma**

| Controls (ng/mL) | Final volume (mL) | µL of 10 µg/mL menthol glucuronide standard | Free menthol concentration |
|---|---|---|---|
| 40 | 10 | 40 | 19 |
| 100 | 10 | 100 | 47 |
| 160 | 10 | 160 | 75 |

Free menthol concentration is calculated as [menthol glucuronide concentration × menthol molecular weight (156)/ menthol glucuronide molecular weight (332)]

**2.5 Analytical Equipment and Supplies**

1. Agilent GC/MS 5975C inert XL MSD with Triple Axis Detector (Agilent Technologies, CA).
2. Analytical column: ZB-1MS 15 m×250 μm×0.25 μm (Phenomenex, Torrance, CA).
3. Carrier Gas: Helium.
4. 13×100 mm screw-cap test tubes with Teflon caps.
5. Autosampler vials with glass inserts and crimp caps (P.J. Cobert Associates, Inc., St. Louis, MO).

# 3   Methods

**3.1 Stepwise Procedure**

1. To 0.5 mL calibrator, control, or sample add 20 μL working IS, 25 μL β-glucuronidase, and 10 μL sodium acetate buffer in labeled 13×100 mm test tubes (*see* **Note 2**).
2. Incubate tubes overnight in 37 °C water bath (*see* **Note 3**).
3. Put tubes in a freezer for 10 min (*see* **Note 4**).
4. Add 150 μL 0.4 M phosphate buffer and 0.5 mL methylene chloride. Extract by rocking for 5 min.
5. Centrifuge for 5 min at 10 °C and 2000×*g*.
6. Remove and discard upper aqueous layer.
7. Transfer bottom layer to autosampler vials and inject 2 μL onto GC/MS.

**3.2 Instrument Operating Conditions**

The instrument's operating conditions are given in Table 3.

**3.3 Data Analysis**

1. Data are analyzed using Target Software (Thru-Put Systems, Orlando, FL) or similar software.

**Table 3**
**GC/MS operating conditions**

| | |
|---|---|
| Oven temperature program | 50 °C for 1 min, 10 °C/min to 90 °C for 5.5 min then 40 °C/min to 250 °C for 2 min Run time: 16.5 min |
| Front inlet | Mode: splitless Injection temperature: 250 °C Column pressure: 7.5 psi Purge time: 0.4 min Septum purge flow: 3 mL/min |
| Mass spectrometer | Mode: electron impact at 70 eV Source temperature: 230 °C Tune: autotune |

**Table 4**
**Quantitation and qualifying ions for menthol**

|  | Quantitation ions | Qualifier ions |
|---|---|---|
| Menthol-d4 | 142 | 99, 127 |
| Menthol | 138 | 95, 123 |

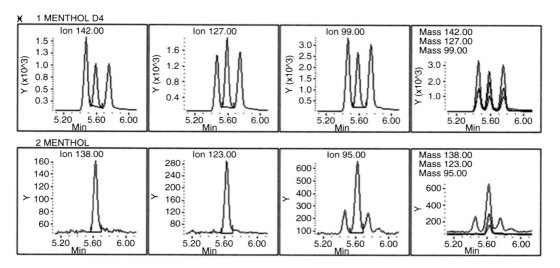

**Fig. 1** Selected ion chromatograms for menthol-d4 and menthol

2. Standard curves are generated based on linear regression of the analyte/IS peak area ratio (y) versus analyte concentration (x) using the quantifying ion listed in Table 4 (*see* **Note 5**). Monitored ions are given in Table 4.

3. Typically, coefficient of correlation is >0.99.

4. Runs are accepted if calculated controls fall within two standard deviations of target values.

5. Within and between run imprecision are <10 %.

6. Representative GC-MS selected ion chromatograms are shown in Fig. 1. Electron impact ionization mass spectra are shown in Figs. 2 and 3.

# 4 Notes

1. It takes ~30 mL phosphate buffer B to adjust the pH to 6.0.

2. Urine samples are analyzed straight and 1:20. Menthol results can be expressed as mg/g creatinine.

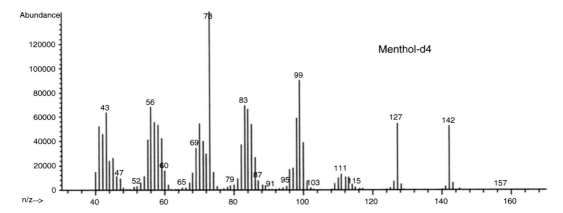

**Fig. 2** Electron impact ionization mass spectrum of menthol-d4

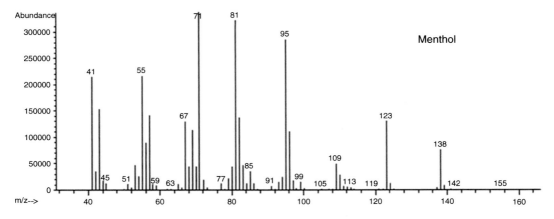

**Fig. 3** Electron impact ionization mass spectrum of menthol

3. Tubes must be tightly capped as menthol is highly volatile.

4. This step is important to keep menthol in liquid phase.

5. Internal standard chromatographs as a triplet of peaks due to mixture of different diastereomers. The middle peak is used for quantitation.

### References

1. Liu B, Fan L, Balakrishna S, Sui A, Morris JB, Jordt SE (2013) TRPM8 is the principal mediator of menthol-induced analgesia of acute and inflammatory pain. Pain 154:2169–2177

2. Farco JA, Grundmann O (2013) Menthol—pharmacology of an important naturally medicinal "cool". Mini Rev Med Chem 13:124–131

3. Kamatou GP, Vermaak I, Viljoen AM, Lawrence BM (2013) Menthol: a simple monoterpene with remarkable biological properties. Phytochemistry 96:15–25

4. Pittler MH, Ernst E (1998) Peppermint oil for irritable bowel syndrome: a critical review and metaanalysis. Am J Gastroenterol 93:1131–1135

5. Ford AC, Talley NJ, Spiegel BM, Foxx-Orenstein AE, Schiller L, Quigley EM, Moayyedi P (2008) Effect of fibre, antispasmodics, and

peppermint oil in the treatment of irritable bowel syndrome: systematic review and meta-analysis. BMJ 337:a2313

6. Rozza AL, Hiruma-Lima CA, Takahira RK, Padovani CR, Pellizzon CH (2013) Effect of menthol in experimentally induced ulcers: pathways of gastroprotection. Chem Biol Interact 206:272–278

7. Gelal A, Jacob P 3rd, Yu L, Benowitz NL (1999) Disposition kinetics and effects of menthol. Clin Pharmacol Ther 66:128–135

8. Mascher H, Kikuta C, Schiel H (2001) Pharmacokinetics of menthol and carvone after administration of an enteric coated formulation containing peppermint oil and caraway oil. Arzneimittelforschung 51:465–469

9. Hamasaki K, Kato K, Watanabe T, Yoshimura Y, Nakazawa H, Yamamoto A, Matsunaga A (1998) Determination of l-menthol in pharmaceutical products by high performance liquid chromatography with polarized photometric detection. J Pharm Biomed Anal 16:1275–1280

10. Kaffenberger RM, Doyle MJ (1990) Determination of menthol and menthol glucuronide in human urine by gas chromatography using an enzyme-sensitive internal standard and flame ionization detection. J Chromatogr 527:59–66

11. Krzek J, Czekaj JS, Rzeszutko W (2003) Validation of a method for simultaneous determination of menthol and methyl salicylate in pharmaceuticals by capillary gas chromatography with cool on-column injection. Acta Pol Pharm 60:343–349

12. Pauwels J, D'Autry W, Van den Bossche L, Dewever C, Forier M, Vandenwaeyenberg S, Wolfs K, Hoogmartens J, Van Schepdael A, Adams E (2012) Optimization and validation of liquid chromatography and headspace-gas chromatography based methods for the quantitative determination of capsaicinoids, salicylic acid, glycol monosalicylate, methyl salicylate, ethyl salicylate, camphor and l-menthol in a topical formulation. J Pharm Biomed Anal 60:51–58

13. Schulz K, Bertau M, Schlenz K, Malt S, Dressler J, Lachenmeier DW (2009) Headspace solid-phase microextraction-gas chromatography-mass spectrometry determination of the characteristic flavourings menthone, isomenthone, neomenthol and menthol in serum samples with and without enzymatic cleavage to validate post-offence alcohol drinking claims. Anal Chim Acta 646:128–140

14. Spichiger M, Muhlbauer RC, Brenneisen R (2004) Determination of menthol in plasma and urine of rats and humans by headspace solid phase microextraction and gas chromatography—mass spectrometry. J Chromatogr B Analyt Technol Biomed Life Sci 799:111–117

# Chapter 23

# Development of an Assay for Methotrexate and Its Metabolites 7-Hydroxy Methotrexate and DAMPA in Serum by LC-MS/MS

Ryan C. Schofield, Lakshmi V. Ramanathan, Kazunori Murata, Martin Fleisher, Melissa S. Pessin, and Dean C. Carlow

## Abstract

Methotrexate (MTX) is a folic acid antagonist that is widely used as an immunosuppressant and chemotherapeutic agent. After high-dose administration of MTX serum levels must be monitored to determine when to administer leucovorin, a folic acid analog that bypasses the enzyme inhibition caused by MTX and reverses its toxicity. We describe a rapid and simple turbulent flow liquid chromatography (TFLC) method implementing positive heated electrospray ionization (HESI) for the accurate and precise determination of MTX, 7-hydroxymethotrexate (7-OH MTX), and 4-amino-4-deoxy-$N^{10}$-methylpteroic acid (DAMPA) concentrations in serum. MTX is isolated from serum samples (100 µL) after protein precipitation with a methanolic solution containing internal standard (MTX-$D_3$) followed by centrifugation. The supernatant is injected into the turbulent flow liquid chromatography which is followed by electrospray positive ionization tandem mass spectrometry (TFLC-ESI-MS/MS) and quantified using a six-point calibration curve. For MTX, 7-OH MTX, and DAMPA the assays were linear from 20 to 1000 nmol/L. Dilutions of 10-, 100-, and 1000-fold were validated giving a clinically reportable range of 20 to $1.0 \times 10^6$ nmol/L. Within-day and between-day precisions at concentrations spanning the analytical measurement ranges were less than 10 % for all three analytes.

**Key words** Methotrexate, Carboxypeptidase-G2, Therapeutic drug monitoring, Mass spectrometry, Turbulent flow liquid chromatography

## 1 Introduction

Methotrexate (MTX) is a folic acid antagonist that is widely used as an immunosuppressant and chemotherapeutic agent. MTX exerts its cytotoxic effects by competitively inhibiting dihydrofolate reductase (DHFR), the enzyme responsible for converting folates to tetrahydrofolates; the folate carrier that functions in the transfer of carbon units. A normal dividing cell uses large amounts of reduced folates to maintain ongoing purine and thymidine synthesis and the demand is even greater for rapidly dividing malignant cells [1].

Uttam Garg (ed.), *Clinical Applications of Mass Spectrometry in Drug Analysis: Methods and Protocols*, Methods in Molecular Biology, vol. 1383, DOI 10.1007/978-1-4939-3252-8_23, © Springer Science+Business Media New York 2016

High-dose MTX is mainly used for the treatment of leukemia and osteosarcoma. Intermediate and lower dose MTX regimens are used to treat malignant gestational trophoblastic disease, breast and bladder cancer, ALL, and acute promyelocytic leukemia [1, 2]. In addition to its antiproliferative activity, MTX also has anti-inflammatory and immunomodulating properties and is a first-line treatment for a growing number of autoimmune rheumatologic, dermatologic, and gasteroenterologic conditions [1, 2]. After high-dose administration of MTX serum levels must be monitored to determine when to administer leucovorin, a folic acid analog that bypasses the enzyme inhibition caused by MTX and reverses its toxicity [3]. Patients in renal failure who are given high-dose MTX are sometimes given carboxypeptidase-G2 $(CPDG_2)$ $(CPDG_2)$ to reverse the effects of MTX [3–5]. $CPDG_2$ is an enzyme that converts MTX into glutamate and 4-amino-4-deoxy-N10-methylpteroic acid (DAMPA) that are much less toxic and readily excreted. DAMPA cross-reacts considerably in immunoassays rendering them unsuitable for monitoring patients who have been given $CPDG_2$ therapy [6].

MTX assays using mass spectrometry have been described previously [7–9]. However the objective of this study was to develop an MTX assay performed by LC-MS with the following characteristics: analytically sensitive with a clinically useful dynamic range; good specificity with no interference from metabolites or other compounds; suitable analytical transferability for a high volume clinical laboratory, and the accurate measurement of 7-OH MTX and DAMPA to support clinical trials utilizing $CPDG_2$ and related compounds. The following chapter describes a rapid and simple turbulent flow method implementing positive heated electrospray ionization for the accurate and precise determination of MTX, 7-OH MTX, and DAMPA concentrations in serum.

# 2    Materials

## 2.1    Samples

Serum samples are required. All samples should be processed and analyzed within 4 h of collection or refrigerated for analysis up to 24 h after collection or frozen for analysis up to 6 months.

## 2.2    Solvents and Reagents

1. Human drug-free pooled normal serum (UTAK Laboratories).

2. Mobile Phase A (10 mM ammonium formate/0.1 % formic acid in water): Remove 8.4 mL of water from a 4 L bottle. Add 2.8 mL of ammonium hydroxide, cap, and invert ten times. Add 5.6 mL of formic acid and degas for 5 min by sonication. The mobile phase is stable at room temperature, 18–24 °C, up to 1 month.

3. Mobile phase B (10 mM ammonium formate/0.1 % formic acid in methanol): Remove 8.4 mL of methanol from a 4 L bottle. Add 2.8 mL of ammonium hydroxide, cap, and invert ten times. Add 5.6 mL of formic acid and degas for 5 min by sonication. The mobile phase is stable at room temperature, 18–24 °C, up to 1 month.

4. Mobile phase C (acetonitrile/2-propanol/acetone, 6:3:1): In a 1000 mL graduated cylinder, add 600 mL acetonitrile, 300 mL of 2-propanol, and 100 mL of acetone into a 2 L HPLC solvent bottle. Degas the solution for 5 min by sonication. The mobile phase is stable at room temperature, 18–24 °C, up to 1 month.

5. Autosampler aqueous wash (water/acetic acid/acetonitrile, 8.8:1:0.2): In a 500 mL graduated cylinder, add 440 mL of water, 50 mL of acetic acid, and 10 mL of acetonitrile and transfer into an HPLC wash bottle. Degas the solution for 5 min by sonication. The wash solution is stable at room temperature, 18–24 °C, up to 1 month.

6. Autosampler organic wash (acetonitrile/2-propanol/acetone, 6:3:1): In a 1000 mL graduated cylinder, add 600 mL acetonitrile, 300 mL of 2-propanol, and 100 mL of acetone and transfer into a 2 L HPLC solvent bottle. Degas the solution for 5 min by sonication. The mobile phase is stable at room temperature, 18–24 °C, up to 1 month.

7. Extraction solution (30 ng/mL MTX-D$_3$ in methanol with 0.1 % formic acid): Add approximately 150 mL of methanol and 200 μL formic acid into a 200 mL volumetric flask. Add 60 μL of MTX-D$_3$ stock (100 μg/mL) to the same volumetric flask. Bring to volume with methanol and mix well. The extraction solution is stable for up to 6 months when stored at −20 °C.

### 2.3  Internal Standards and Standards

1. Primary standards: MTX (Sigma-Aldrich), 7-OH MTX (Santa Cruz Biotechnology Inc.), and DAMPA (Schircks Laboratories).

2. Primary internal standard (I.S.): MTX-D$_3$ (Cerilliant) 100 μg/mL in 0.1 N sodium hydroxide.

3. Standard stock solutions:

   (a) Methotrexate MTX (1 mg/mL): Using an analytical balance weigh 25 mg and place in a 25 mL volumetric flask. Bring to volume with methanol containing 0.1 N sodium hydroxide and mix well. Then prepare a 100 μg/mL stock solution from the previous 1 mg/mL stock solution: Add 900 μL of methanol to a 2 mL amber vial and add 100 μL of MTX stock (1 mg/mL) and mix well. Both stock solutions are stable up to 6 months when stored at −20 °C.

(b) 7-Hydroxymethotrexate, 7-OH MTX (1 mg/mL): Using an analytical balance weigh 25 mg and place in a 25 mL volumetric flask. Bring to volume with methanol containing 0.1 N sodium hydroxide and mix well. Then prepare a 100 µg/mL stock solution from the previous: Add 900 µL of methanol to a 2 mL amber vial and add 100 µL of 7-OH MTX stock (1 mg/mL) and mix well. Both stock solutions are stable for up to 6 months when stored at −20 °C.

(c) DAMPA (1 mg/mL): Using an analytical balance weigh 25 mg and place in a 25 mL volumetric flask. Bring to volume with methanol containing 0.1 N sodium hydroxide and mix well. Then prepare a 100 µg/mL stock solution from the previous: Add 900 µL of methanol to a 2 mL amber vial and add 100 µL of DAMPA stock (1 mg/mL) and mix well. Both stock solutions are stable for up to 6 months when stored at −20 °C.

(d) Combined standard (MTX, 7-OH MTX, and DAMPA): To obtain a combined standard containing, 4.54 µg/mL, 9.40 µg/mL, and 3.25 µg/mL respectively, add 1656 µL of drug-free serum to a 2 mL amber glass vial. Then add 91, 188, and 65 µL of MTX, 7-OH MTX, and DAMPA from each respective 100 µg/mL stock solution and mix well. Stock solution is stable for up to 6 months when stored at −20 °C.

## 2.4 Calibrators and Controls

1. Calibrators: Prepare calibrators 1–8 by making serial dilutions of the combined standard according to Table 1. For each dilution step add the appropriate amount of previous solution as shown in the table to a 10.0 mL volumetric flask and fill with drug-free human serum. Vortex mix the volumetric flask after each dilution step (*see* **Note 1**).

2. Controls: MTX controls were purchased from UTAK laboratories at the following concentrations: 0.023, 0.034, 0.227, and 0.341 µg/mL. Currently there are no commercially available controls for 7-OH MTX and DAMPA. To prepare 7-OH MTX and DAMPA controls follow Table 2 for the procedure. For each control add the appropriate amount of 100 µg/mL stock solution into a 10 mL volumetric flask then bring to volume with drug-free serum. These two sets of controls are made on separate days and from separate lots of material than the calibrators (*see* **Note 1**).

3. Check the new lot of standards by verifying five unknown patient samples concentrations with the current lot of calibrators. The agreement between the two calculated concentrations must be within 10 %.

4. Establish a range for the new lot of controls by collecting data points over 20 consecutive runs and establish the mean and standard deviation.

**Table 1**
**Calibrator preparation**

| Calibrator | Volume of previous standard (mL) | Drug-free serum (mL) | Final concentrations (µg/mL) MTX 7-OH MTX DAMPA | | |
|---|---|---|---|---|---|
| 1 | 1.0 (combined std.) | 9.0 | 0.454 | 0.940 | 0.325 |
| 2 | 5.0 | 5.0 | 0.227 | 0.470 | 0.163 |
| 3 | 5.0 | 5.0 | 0.114 | 0.235 | 0.081 |
| 4 | 5.0 | 5.0 | 0.057 | 0.118 | 0.041 |
| 5 | 5.0 | 5.0 | 0.028 | 0.059 | 0.020 |
| 6 | 5.0 | 5.0 | 0.014 | 0.029 | 0.010 |
| 7 | 5.0 | 5.0 | 0.007 | 0.015 | 0.005 |
| 8 | 5.0 | 5.0 | 0.004 | 0.007 | 0.003 |

*MTX*: 1 µg/mL = 2201.8 nmol/L
*7-OH MTX*: 1 µg/mL = 2127.1 nmol/L
*DAMPA*: 1 µg/mL = 3075.7 nmol/L

**Table 2**
**7-OH MTX and DAMPA control preparation**

| Control | Volume of 7-OH MTX 100 µg/mL stock | Volume of DAMPA 100 µg/mL stock | Final concentration (µg/mL) 7-OH MTX DAMPA | |
|---|---|---|---|---|
| Low | 24 µL | 8 µL | 0.235 | 0.081 |
| Mid | 47 µL | 16 µL | 0.470 | 0.161 |
| High | 71 µL | 24 µL | 0.705 | 0.244 |

**2.5 Analytical Equipment and Supplies**

1. Thermo Scientific Transcend TLX-2 with Agilent 1200 pumps coupled to a TSQ Quantum Ultra triple quadrupole mass spectrometer running Aria software 1.6.3 (Thermo Scientific).

2. TurboFlow column (TFC): Cyclone-P 50×0.5 mm, 60 µm particle size, 60 Å pore size (Thermo Scientific). Analytical column: Hypersil Gold C8 2.1×50 mm, 5 µm particle size (Thermo Scientific).

3. Column heater (Thermo Scientific).

4. Eppendorf 1.5 mL microcentrifuge tubes and National Scientific 2 mL amber glass vials with inserts and pre-slit caps or equivalent.

**2.6 Instrument Operating Conditions**

1. Turbulent flow liquid chromatography (TFLC): Chromatographic separations were performed using a Thermo Scientific Transcend TLX-2 which was comprised of a PAL autosampler (CTC Analytics), a low-pressure mixing quaternary pump (loading pump), a high-pressure binary pump

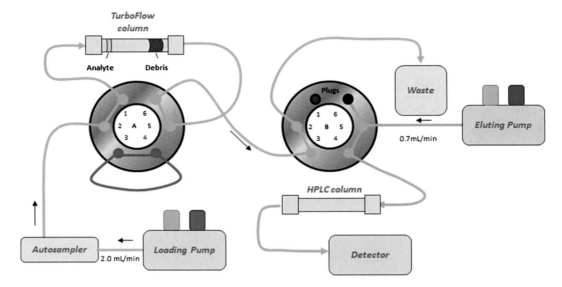

**Fig. 1** Transcend TLX-2 valve configuration

(eluting pump), and a six-valve switching module with six-port valves (Fig. 1). The system was controlled via Aria software, version 1.6.2. The TurboFlow column used was a Cyclone-P 50×0.5 mm, 60 μm particle size, and a 60 Å pore size. The analytical HPLC column was a Hypersil Gold C8 2.1×50 mm, 5 μm particle size. The temperature of the analytical column was maintained at 70 °C using a column heater. The analytes were loaded on the TurboFlow column in 100 % mobile phase A and transferred to the HPLC column with 80 % mobile phase B using a 200 μL transfer loop. The loading and eluting mobile phase composition for the HPLC column was identical to that of the TurboFlow column. The integration parameters for all four analytes were similar with a baseline window of 20, area noise factor of 5, peak noise factor of 10, and an integration window of 15 s. The retention times of the TFLC-ESI-MS/MS ion chromatograms of the analytes can be seen in Fig. 2.

2. Tandem mass spectrometry: Mass spectrometric detection was performed using a Thermo Scientific TSQ Quantum Ultra triple-quadrupole mass spectrometer equipped with an electrospray ionization (ESI) source operating in a positive ion mode. MS/MS conditions are depicted in Table 4. Nitrogen (99.995 % purity) was used as the desolvation gas, and ultrapure argon (99.999 % purity) was used as the collision gas. The mass transitions from the protonated molecular ion $[M+H]^+$ to the most abundant product ions were used as the quantifying ions for each analyte (Table 5). The SRM acquisition method was run in unit resolution (0.7) in both Q1 and Q3 with a scan width and scan rate of 0.050 m/z and 0.100 s, respectively.

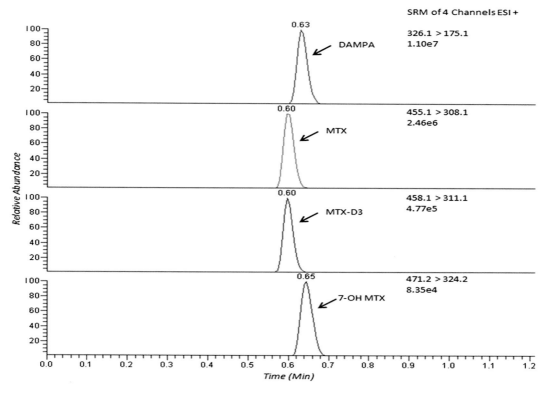

**Fig. 2** TFLC-ESI-MS/MS ion chromatograms of MTX, 7-OH MTX, DAMPA, and MTX-D$_3$ (I.S.) product ions

## 3   Methods

**3.1   Stepwise Procedure**

1. Run a system suitability to confirm the system performance (*see* **Note 2**).

2. To a 1.5 mL microcentrifuge tubes pipette 100 μL serum (calibrators, controls, or patient specimen) (*see* **Note 3**).

3. Add 200 μL extraction solution.

4. Cap and vortex each sample vigorously.

5. Centrifuge for 10 min at 13,000×*g*.

6. Transfer 200 μL of the supernatant into glass inserts and place in 2 mL amber glass vials.

7. Place all samples in the autosampler and inject 50 μL of the sample into the TFLC-ESI-MS/MS. Ion chromatograms for all analytes are shown in Fig. 2.

**Table 3**
**TurboFlow and HPLC operating parameters**

| A. HPLC method. TurboFlow parameters HPLC parameters | | | | | | | | | | | | | |
|---|---|---|---|---|---|---|---|---|---|---|---|---|---|
| Step | Start | s | Flow | Grad | %A | %B | %C | Tee | Loop | Flow | Grad | %A | %B |
| 1 | 0:00 | 30 | 2.00 | Step | 100 | | | ==== | Out | 0.7 | Step | 100 | |
| 2 | 0:30 | 45 | 0.15 | Step | 100 | | | T | In | 0.7 | Step | 100 | |
| 3 | 1:15 | 15 | 2.00 | Step | | | 100 | ==== | In | 0.7 | Ramp | 20 | 80 |
| 4 | 1:30 | 15 | 2.00 | Step | | | 100 | ==== | Out | 0.7 | Step | 20 | 80 |
| 5 | 1:45 | 30 | 2.00 | Step | | | 100 | ==== | In | 0.7 | Step | 20 | 80 |
| 6 | 2:15 | 45 | 2.00 | Step | 20 | 80 | | ==== | Out | 0.7 | Step | 20 | 80 |
| 7 | 3:00 | 45 | 2.00 | Step | 20 | 80 | | ==== | In | 0.7 | Step | | 100 |
| 8 | 3:45 | 75 | 2.00 | Step | 100 | | | ==== | Out | 0.7 | Step | 100 | |

| B. Mobile phase composition | |
|---|---|
| Mobile phase | |
| Loading pump A: | 10 mM ammonium formate/0.1 % formic acid in water (v/v) |
| Loading pump B: | 10 mM ammonium formate/0.1 % formic acid in methanol (v/v) |
| Loading pump C: | Acetonitrile/2-propanol/acetone (6:3:1 v/v) |
| Eluting pump A: | 10 mM ammonium formate/0.1 % formic acid in water (v/v) |
| Eluting pump B: | 10 mM ammonium formate/0.1 % formic acid in methanol (v/v) |
| Aqueous wash 1: | Water/acetic acid/acetonitrile (8.8:1:0.2 v/v) |
| Organic wash 2: | Acetonitrile/2-propanol/acetone (6:3:1 v/v) |

**3.2 Analysis**

1. Instrumental operating parameters are shown in Tables 3 and 4.

2. The data are analyzed using LCquan 2.6 software (Thermo Scientific).

3. Standard curves are based on linear regression analysis for MTX, 7-OH MTX, and DAMPA. Weighted linear regression models with weights inversely proportional to the X values were used. The analysis compared I.S. peak area to sample peak area (y-axis) versus analyte concentration (x-axis) using the quantifying ions indicated in Table 5.

**Table 4**
**MS/MS tune settings**

| | |
|---|---|
| Spray voltage (V): | 4500 |
| Vaporizer temperature (°C): | 380 |
| Sheath gas pressure (arbitrary units): | 60 |
| Ion sweep gas pressure (arbitrary units): | 2.0 |
| Aux gas pressure (arbitrary units): | 15 |
| Capillary temperature (°C): | 235 |
| Collision pressure (mTorr): | 1.5 |
| Data window (min): | 1:45–3:15 |

**Table 5**
**Analyte precursor and product ions**

| Analyte | Precursor ion (*m/z*) | Product ion (*m/z*) | CE (V) | Tube lens (V) | Skimmer (V) | CFPW (s) |
|---|---|---|---|---|---|---|
| MTX | 455.1 | 308.1 | 18 | 105 | 10 | 5 |
| 7-OH MTX | 471.2 | 324.2 | 11 | 105 | 10 | 5 |
| DAMPA | 326.1 | 175.1 | 18 | 110 | 10 | 5 |
| MTX-D3 | 458.1 | 311.1 | 18 | 105 | 10 | 5 |

Optimized m/z may change based on instrument and tuning parameters

4. Acceptability of each run is confirmed if the calculated control concentrations fall within two standard deviations of the target mean values. Target values are established as the mean of 20 separate runs. If any control is greater than three standard deviations from the mean the run cannot proceed and troubleshooting procedure must commence.

5. Typical coefficients of correlation of the standard curve are >0.995 (*see* **Note 4**).

# 4   Notes

1. Individual sets of calibrators and controls can be aliquoted and stored at −80 °C for 1 year. For each calibrator or control, aliquot 125 µL into a glass insert and place in a 2 mL amber glass vial and cap. Thaw completely before use.

2. A system suitability should be performed each day the method is run. The suitability includes running a test mix with all analytes to ensure proper retention time, integration, and sensitivity.

3. A new standard curve should be generated with each analytical run to ensure method performance.

4. The MTX, 7-OH MTX, and DAMPA assays were linear from 0 to 0.454 µg/mL, 0.007 to 0.940 µg/mL, and 0.003 to 0.325 µg/mL, respectively. Dilutions of 10-, 100-, and 1000-fold were validated for all analytes. Within-day and between-day precisions at concentrations spanning the analytical measurement ranges were less than 10 % for all three analytes.

## References

1. LaCasce AS (2014) Therapeutic use and toxicity of high-dose methotrexate. In: Post TW (ed) UpToDate. Waltham, MA. Accessed 1 Oct 2014

2. Lennard L (1999) Therapeutic drug monitoring of antimetabolic cytotoxic drugs. Br J Clin Pharmacol 47:131–143

3. Widemann BC, Adamson PC (2006) Understanding and managing methotrexate nephrotoxicity. Oncologist 11:694–703

4. DeAngelis LM, Tong WP, Lin S, Fleisher M et al (1996) Carboxypeptidase G2 rescue after high-dose methotrexate. J Clin Oncol 14:2145–2149

5. Widemann BC, Sung E, Anderson L et al (2000) Pharmacokinetics and metabolism of the methotrexate metabolite 2,4-diamino-N-methylpteroic acid. J Pharmacol Exp Ther 294:894–901

6. Albertioni F, Rask C, Eksborg S et al (1996) Evaluation of clinical assays for measuring high-dose methotrexate in plasma. Clin Chem 42:39–44

7. Kumar VS, Law T, Kellogg M (2010) Liquid chromatography-tandem mass spectrometry (LC-MS-MS) method for monitoring methotrexate in the setting of carboxypeptidase-G2 therapy. In: Garg U, Hammett-Stabler CA (eds) Clinical applications of mass spectrometry: methods in molecular biology 603. Humana Press, New York

8. Guo P, Wang X, Liu L et al (2007) Determination of methotrexate and its major metabolite 7-hydroxymethotrexate in mouse plasma and brain tissue by liquid chromatography-tandem mass spectrometry. J Pharm Biomed Anal 43:1789–1795

9. Nair H, Lawrence L, Hoofnagle AN (2012) Liquid chromatography-tandem mass spectrometry work flow for parallel quantification of methotrexate and other immunosuppressants. Clin Chem 58:943–945

# Chapter 24

# Quantitative, Multidrug Pain Medication Testing by Liquid Chromatography: Tandem Mass Spectrometry (LC-MS/MS)

## Geza S. Bodor

## Abstract

Chronic pain is often treated with narcotic analgesics. The most commonly used narcotic analgesics are the opiates (natural or modified compounds of the poppy plant) or opioids (synthetic chemicals that act on opiate receptors). While opiates and opioids are excellent analgesics, they can also have significant side effects that include respiratory depression, coma, or death. Tolerance, physical dependence, and addiction (psychological dependence) are other severe side effects of opioid use. Patients who develop dependence or addiction often times abuse other, non-opioid narcotics and may trade their prescription medication for illegal street drugs (called "diversion"). In order to minimize side effects, detect possible multidrug abuse and prove diversion, simultaneous monitoring of numerous prescription and illicit drugs is required.

The method described in this chapter is for the quantitative measurement of 43 different drugs in urine. The panel includes narcotic pain medications, benzodiazepines, NIDA drugs, and other, commonly abused medications. The analytes of interests are injected in the presence of deuterated internal standards to correct for possible extraction inefficiencies, ion suppression, or other interferences. The sample is prepared by adding dilution buffer with the deuterated internal standards to the sample, followed by reversed-phase, gradient HPLC separation on a Phenyl-Hexyl column using water and methanol as mobile phases. Detection of the analytes of interest is done by isotope-dilution mass spectrometry on a triple-quadrupole tandem mass spectrometer following electrospray ionization in the positive mode. Mass spectrometric (MS) data are collected in the scheduled MRM (sMRM) mode. Two MRM transitions are monitored for each analyte and one MRM transition is monitored for each IS. Quantitation of the unknown analytes is achieved by comparing the peak area ratios of the analytes to that of the internal standards and reading the unknown concentration from a seven-point calibration curve.

**Key words** Opiates, Opioids, Narcotic analgesic, Drugs of abuse, Liquid chromatography, Tandem mass spectrometry

## 1 Introduction

Approximately 30 % of the US population (~100 million people) are affected by chronic pain [1, 2]. Those who seek help may receive drug therapy for their condition that consists of either NSAIDs or of narcotic analgesics according to the WHO 3-step analgesic protocol [3]. The most commonly used narcotic

Uttam Garg (ed.), *Clinical Applications of Mass Spectrometry in Drug Analysis: Methods and Protocols*, Methods in Molecular Biology, vol. 1383, DOI 10.1007/978-1-4939-3252-8_24, © Springer Science+Business Media New York 2016

analgesics are the opiates (natural or modified compounds of the poppy plant) or opioids (synthetic chemicals that act on opiate receptors). While opiates and opioids are excellent analgesics, they can also have significant side effects that include respiratory depression, coma, or death. Tolerance, physical dependence, and addiction (psychological dependence) are other severe side effects of opioid use. Tolerance and physical dependence, when develop, will require successively higher doses of the drug to achieve the desired analgesic effect. Addiction, on the other hand, is characterized by drug seeking behavior that can lead to criminal activity to obtain the drug. Additional signs of addiction are dysfunctional opioid use and the concurrent use of more than one narcotics, including illicit ones. Previously acquired tolerance can be lost if the opioid drug is discontinued even for a few days. Resuming treatment after a hiatus, at the previously tolerated dose, could lead to opioid overdose with occasional fatal outcome.

To prevent fatal drug overdose, to monitor appropriate medication use, and to assess the possibility of addiction, laboratory monitoring of chronic pain management must use appropriate analytical methods. These methods must be sensitive enough to detect the presence of prescribed drugs even when taken at low dosage or intermittently, as during "as needed" administration. The analytical method must also exhibit sufficient analytical specificity for the individual analytes of interest to be able to detect possible concurrent use of similar drugs such as morphine and hydromorphone or morphine and codeine. Patients who are addicted to narcotic pain medications often use illicit drugs; therefore, they must be monitored for illegal drug use as well.

When we test for pain medications we assess the patient's compliance with his/her prescription and we look for the presence of non-prescribed or illicit compounds. Confirmed absence of the prescribed drug is interpreted as "diversion," or the illegal sale of prescription medication, while the presence of non-prescribed drugs may be a sign of addiction. Either one of these occurrences can lead to discharge of the patient from the treatment program. Compliance with treatment guidelines is only proven if the patient has the prescribed drug(s) or their metabolite(s) in his/her body while he/she does not have any non-prescribed narcotics on board.

These multiple requirements for pain medication monitoring mandate the use of chromatography-mass spectrometry-based methods for the desired sensitivity and specificity, and they also require the use of panels that can simultaneously detect and quantitate the most frequently used and abused drugs in a single sample [4].

This is a panel for the quantitative measurement and reporting of 43 different drugs in urine. The panel includes narcotic pain medications, benzodiazepines, NIDA drugs, and other, commonly abused medications. The analytes of interests are injected in the presence of deuterated internal standards to correct for possible extraction inefficiencies, ion suppression, or other interferences.

The sample preparation method is a so-called "dilute and shoot" method. It is based on reversed-phase HPLC separation and isotope-dilution mass spectrometry analysis on a triple-quadrupole tandem mass spectrometer following electrospray (a.k.a. TurboIonSpray or TIS) ionization in the positive mode. Mass spectrometry scan is performed in the sMRM mode. Each unknown analyte is identified by two, specific transitions and each deuterated IS is identified by a single transition.

Quantitation of the unknown analytes is achieved by comparing the peak area ratios of the analyte to that of the internal standard and reading the unknown concentration from a seven-point calibration curve.

# 2  Materials

## 2.1  Samples

Timed or random urine are acceptable specimens for this testing. Samples are stable for 7 days at room temperature, 30 days refrigerated, or 12 months frozen.

## 2.2  Reagents

1. Drug-free urine (Utak Laboratories).

2. Multi-constituent Pain Panel quality control at three different concentrations, custom ordered (Utak Laboratories).

3. Mobile Phase A/Buffer A (10 mM Ammonium Formate in water): To 1.0 L LC/MS Optima grade water add 1 mL of 10 M Ammonium Formate. Mix well. Store at room temperature on instrument. Reagent expires after 3 days and must be discarded if not used within 72 h.

4. Mobile Phase B/Buffer B (Methanol with 0.1 % Formic Acid): To 500 mL of LC/MS Optima grade methanol add 500 μL formic acid. Store at room temperature (18–26 ° C) on instrument. Reagent expires after 2 days and must be discarded. Extreme high room temperature will accelerate deterioration of mobile phase B. If room temperature is above 26 ° C this buffer must be made daily (*See* **Note 1**).

5. Sample Diluent (SD): Mix together 180 mL Buffer A and 20 mL Buffer B. Mix well. Sample diluent is used to prepare IS Working solution. Store refrigerated (2–8 °C) in tightly closed container. Reagent expires 7 days from preparation.

6. Internal Standard Stock Solution (IS Stock): Pipette the volume of Cerilliant deuterated stock listed in column "Spike volume (μL)" in Table 1. into a 10 mL volumetric flask. QS the content of flask to 10 mL with MeOH. Store IS Stock solution in –70 ° C freezer. Stable for 6 months.

**Table 1**
**Preparation of IS stock solution**

| Deuterated chemical | Cerilliant catalogue number | Cerilliant stock conc (µg/mL) | Spike volume (µL) | IS Stock solution conc (ng/mL) | IS Working solution conc (ng/mL) |
|---|---|---|---|---|---|
| 6-Acetylmorphine-D3 (6-MAM-D3) | A-006 | 100 | 60 | 600 | 60 |
| Amphetamine-D5 | A-005 | 100 | 400 | 4000 | 400 |
| Benzoylecgonine-D8 | B-001 | 100 | 500 | 5000 | 500 |
| Codeine-D6 | C-040 | 100 | 500 | 5000 | 500 |
| Fentanyl-D5 | F-001 | 100 | 20 | 200 | 20 |
| Hydrocodone-D6 | H-047 | 100 | 500 | 5000 | 500 |
| Hydromorphone-D6 | H-049 | 100 | 500 | 5000 | 500 |
| Meperidine-D4 | M-036 | 100 | 500 | 5000 | 500 |
| Methadone-D3 | M-008 | 100 | 500 | 5000 | 500 |
| Methamphetamine-D5 | M-004 | 100 | 500 | 5000 | 500 |
| Morphine-D6 | M-085 | 100 | 500 | 5000 | 500 |
| Nordiazepam-D5 | N-903 | 100 | 100 | 1000 | 100 |
| Norbuprenorphine-D3 | N-920 | 100 | 250 | 2500 | 250 |
| Norpropoxyphene maleate-D5 | N-904 | 100 | 500 | 5000 | 500 |
| PCP-D5 | P-003 | 100 | 100 | 1000 | 100 |
| (±)-11-Hydroxy-.9-THC-D3 (THC-OH-D3) | H-041 | 100 | 100 | 1000 | 100 |
| (±)-11-nor-9-Carboxy-.9-THC-D3 (THC-COOH-D3) | T-004 | 100 | 100 | 1000 | 100 |
| Tramadol-C13-D3 | T-029 | 100 | 200 | 2000 | 200 |

7. Internal Standard Working Solution (IS Working): Add 2.0 mL of IS Stock solution to 100 mL Sample Diluent. Store IS Working solution refrigerated (2–8 ° C). Reagent expires when Sample Diluent used for preparation expires.

**2.3 Deuterated Internal Standards, Standards, and Calibrators**

1. All deuterated internal stock solutions and standards were purchased from Cerilliant, Inc. (www.cerilliant.com). The list of chemicals and the manufacturer's catalog numbers to be used for the preparation of working internal standards (IS) are presented in Table 1. Chemicals are stable unopened as stated by the manufacturer.

2. Fentanyl Sub-Stock solution: Using the Cerilliant 1000 ng/mL Fentanyl stock solution (F-013), prepare 50 ng/mL sub-stock solution by adding 50 µL Cerilliant Fentanyl Stock to 950 µL MeOH. Mix well. Store frozen at –70 °C. Solution expires with original Cerilliant stock expiration date.

3. Calibrator Stock (100×) solution: Prepare Calibrator Stock (100×) solution by pipetting the amount indicated in the "Spike volume (µL)" column of Table 2 into a 10 mL volumetric flask. QS the content of the flask to 10 mL with methanol. Mix well. Aliquot 1.1 mL of the Calibrator Stock (100×) solution into labeled 4 mL autosampler vials. Store at –70 C for up to 1 year. *See* **Note 2**.

4. High Calibrator (Calibrator G) solution: Pipette 1.0 mL of Calibrator Stock (100×) solution into a 10 mL flask. QS to 10 mL with Drug Free Urine. The High Calibrator (Calibrator G) solution is the highest concentration calibrator in the calibration curve. Use Calibrator G immediately to prepare the other calibrators of the multi-constituent Pain Panel calibrators as described below.

5. Preparation of Calibrators A through F for the Pain Panel Calibration Curve: Levels for a seven-point calibration curve are prepared using the High Calibrator (Calibrator G) solution and Drug-Free urine according to Table 3. After preparation aliquot 500 µL of each solution into 4 mL autosampler vials. Label and store aliquots frozen in the –70 °C freezer. Unopened expiration is 1 year. Store opened vials refrigerated for up to 1 month.
   The nominal concentrations of each drug in the individual Calibrators are listed in Table 4. (*See* **Note 3**).

*2.4 Supplies and Equipment*

1. Kinetex Phenyl-Hexyl HPLC column, 50 mm×4.6 mm×2.6 µm, part # 00B-4495-E0, Phenomenex.

2. SecurityGuard Ultra Cartridges for UHPLC Phenyl 4.6 mm ID column, 3/pack, part # AJ0-8774, Phenomenex.

3. SecurityGuard Ultra Cartridge Holder, part # AJ0-9000, Phenomenex.

4. National Scientific 4 mL autosampler vials, 15×45 mm, 13–425 screw-top vials, convenience kit, cat. # 03-391-7B, Fisher Scientific (or equivalent).

5. 1.5 mL autosampler vial kit, clear glass, 9 mm screw cap, 100/pk (with "one twist" cap and pre-inserted PTFE/silicone septum. Vial size is 12×32 mm, with a writing patch on the side) Shimadzu part number 228-45450-91.

6. 1.0 mL autosampler vial kit, clear glass, 8×40 mm shell vials with PE plugs, borosilicate type I class B glass. 250/pk Shimadzu part# 220-91521-06.

**Table 2**
**Preparation of calibrator stock (100×) solution**

| Analyte | Cerilliant catalogue number | Cerilliant stock conc (µg/mL) | Spike volume (µL) | Calibrator Stock (100×) conc (µg/mL) |
|---|---|---|---|---|
| 6-Acetylmorphine (6-MAM) | A-009 | 1000 | 10 | 1 |
| Alpha-hydroxyalprazolam | A-907 | 1000 | 20 | 2 |
| Alprazolam | A-903 | 1000 | 20 | 2 |
| Amphetamine | A-007 | 1000 | 250 | 25 |
| Benzoylecgonine | B-004 | 1000 | 100 | 10 |
| Buprenorphine | B-902 | 100 | 500 | 5 |
| Carisoprodol | C-077 | 1000 | 300 | 30 |
| Clonazepam | C-907 | 1000 | 20 | 2 |
| Codeine | C-006 | 1000 | 300 | 30 |
| Diazepam | D-907 | 1000 | 20 | 2 |
| EDDP perchlorate | E-022 | 1000 | 200 | 20 |
| Fentanyl (* use the previously prepared 50 µL/mL sub-stock) | F-013 | 50* | 20* | 0.1 |
| Flunitrazepam | F-907 | 1000 | 20 | 2 |
| Flurazepam | F-003 | 1000 | 20 | 2 |
| Hydrocodone | H-003 | 1000 | 100 | 10 |
| Hydromorphone | H-004 | 1000 | 100 | 10 |
| Lorazepam | L-901 | 1000 | 20 | 2 |
| MDA (3,4-Methylenedioxyamphetamine) | M-012 | 1000 | 250 | 25 |
| MDEA (3,4-Methylenedioxyethylamphetamine) | M-065 | 1000 | 250 | 25 |
| MDMA (3,4-Methylenedioxymethamphetamine) | M-013 | 1000 | 250 | 25 |
| Meperidine | M-035 | 1000 | 150 | 15 |
| Meprobamate | M-039 | 1000 | 300 | 30 |
| Methadone | M-007 | 1000 | 200 | 20 |
| Methamphetamine | M-009 | 1000 | 250 | 25 |
| Midazolam | M-908 | 1000 | 20 | 2 |
| Morphine | M-005 | 1000 | 300 | 30 |
| Naloxone | N-004 | 1000 | 150 | 15 |

(continued)

**Table 2**
**(continued)**

| Analyte | Cerilliant catalogue number | Cerilliant stock conc ($\mu$g/mL) | Spike volume ($\mu$L) | Calibrator Stock (100×) conc ($\mu$g/mL) |
|---|---|---|---|---|
| Naltrexone | N-007 | 1000 | 150 | 15 |
| Norbuprenorphine | N-912 | 100 | 500 | 5 |
| Nordiazepam | N-905 | 1000 | 20 | 2 |
| Norfentanyl oxalate | N-031 | 1000 | 5 | 0.5 |
| Normeperidine | N-017 | 100 | 1000 | 10 |
| Norpropoxyphene maleate | N-913 | 1000 | 200 | 20 |
| Oxazepam | O-902 | 1000 | 20 | 2 |
| Oxycodone | O-002 | 1000 | 100 | 10 |
| Oxymorphone | O-004 | 1000 | 100 | 10 |
| Phencyclidine (PCP) | P-007 | 1000 | 25 | 2.5 |
| Propoxyphene | P-011 | 1000 | 200 | 20 |
| Sufentanil citrate | S-008 | 100 | 50 | 0.5 |
| Temazepam | T-907 | 1000 | 20 | 2 |
| (±)-11-Hydroxy-.9-THC (THC-OH) | H-027 | 1000 | 30 | 3 |
| (−)-11-nor-9-Carboxy-.9-THC (THC-COOH) | T-019 | 1000 | 30 | 3 |
| Tramadol HCl | T-027 | 1000 | 100 | 10 |

**Table 3**
**Preparation of the individual calibrators of the multi-constituent calibration curve**

| Calibrator | High calibrator (Calibrator G) (mL) | Drug free urine (mL) | Final volume (mL) | Expected conc (x Cal C) |
|---|---|---|---|---|
| Calibrator G | 4.00 | – | 4.00 | 10 |
| Calibrator F | 2.00 | 2.00 | 4.00 | 5 |
| Calibrator E | 1.00 | 3.00 | 4.00 | 2.5 |
| Calibrator D | 0.50 | 3.50 | 4.00 | 1.25 |
| Calibrator C | 0.40 | 3.60 | 4.00 | 1 |
| Calibrator B | 0.30 | 3.70 | 4.00 | 0.75 |
| Calibrator A | 0.20 | 3.80 | 4.00 | 0.5 |

**Table 4**
**Nominal concentrations of each drug in the individual calibrators**

|  | Cal A | Cal B | Cal C | Cal D | Cal E | Cal F | Cal G |
|---|---|---|---|---|---|---|---|
| 6-MAM | 5.00 | 7.50 | 10.00 | 12.50 | 25.00 | 50.00 | 100.00 |
| Alpha-hydroxyalprazolam | 10.00 | 15.00 | 20.00 | 25.00 | 50.00 | 100.00 | 200.00 |
| Alprazolam | 10.00 | 15.00 | 20.00 | 25.00 | 50.00 | 100.00 | 200.00 |
| Amphetamine | 125.00 | 187.50 | 250.00 | 312.50 | 625.00 | 1250.00 | 2500.00 |
| Benzoylecgonine | 50.00 | 75.00 | 100.00 | 125.00 | 250.00 | 500.00 | 1000.00 |
| Buprenorphine | 25.00 | 37.50 | 50.00 | 62.50 | 125.00 | 250.00 | 500.00 |
| Carisoprodol | 150.00 | 225.00 | 300.00 | 375.00 | 750.00 | 1500.00 | 3000.00 |
| Clonazepam | 10.00 | 15.00 | 20.00 | 25.00 | 50.00 | 100.00 | 200.00 |
| Codeine | 150.00 | 225.00 | 300.00 | 375.00 | 750.00 | 1500.00 | 3000.00 |
| Diazepam | 10.00 | 15.00 | 20.00 | 25.00 | 50.00 | 100.00 | 200.00 |
| EDDP | 100.00 | 150.00 | 200.00 | 250.00 | 500.00 | 1000.00 | 2000.00 |
| Fentanyl | 0.50 | 0.75 | 1.00 | 1.25 | 2.50 | 5.00 | 10.00 |
| Flunitrazepam | 10.00 | 15.00 | 20.00 | 25.00 | 50.00 | 100.00 | 200.00 |
| Flurazepam | 10.00 | 15.00 | 20.00 | 25.00 | 50.00 | 100.00 | 200.00 |
| Hydrocodone | 50.00 | 75.00 | 100.00 | 125.00 | 250.00 | 500.00 | 1000.00 |
| Hydromorphone | 50.00 | 75.00 | 100.00 | 125.00 | 250.00 | 500.00 | 1000.00 |
| Lorazepam | 10.00 | 15.00 | 20.00 | 25.00 | 50.00 | 100.00 | 200.00 |
| MDA | 125.00 | 187.50 | 250.00 | 312.50 | 625.00 | 1250.00 | 2500.00 |
| MDEA | 125.00 | 187.50 | 250.00 | 312.50 | 625.00 | 1250.00 | 2500.00 |
| MDMA | 125.00 | 187.50 | 250.00 | 312.50 | 625.00 | 1250.00 | 2500.00 |
| Meperidine | 75.00 | 112.50 | 150.00 | 187.50 | 375.00 | 750.00 | 1500.00 |
| Meprobamate | 150.00 | 225.00 | 300.00 | 375.00 | 750.00 | 1500.00 | 3000.00 |
| Methadone | 100.00 | 150.00 | 200.00 | 250.00 | 500.00 | 1000.00 | 2000.00 |
| Methamphetamine | 125.00 | 187.50 | 250.00 | 312.50 | 625.00 | 1250.00 | 2500.00 |
| Midazolam | 10.00 | 15.00 | 20.00 | 25.00 | 50.00 | 100.00 | 200.00 |
| Morphine | 150.00 | 225.00 | 300.00 | 375.00 | 750.00 | 1500.00 | 3000.00 |
| Naloxone | 75.00 | 112.50 | 150.00 | 187.50 | 375.00 | 750.00 | 1500.00 |
| Naltrexone | 75.00 | 112.50 | 150.00 | 187.50 | 375.00 | 750.00 | 1500.00 |
| Norbuprenorphine | 25.00 | 37.50 | 50.00 | 62.50 | 125.00 | 250.00 | 500.00 |
| Nordiazepam | 10.00 | 15.00 | 20.00 | 25.00 | 50.00 | 100.00 | 200.00 |
| Norfentanyl | 2.50 | 3.75 | 5.00 | 6.25 | 12.50 | 25.00 | 50.00 |
| Normeperidine | 50.00 | 75.00 | 100.00 | 125.00 | 250.00 | 500.00 | 1000.00 |

(continued)

**Table 4**
**(continued)**

|  | Cal A | Cal B | Cal C | Cal D | Cal E | Cal F | Cal G |
|---|---|---|---|---|---|---|---|
| Norpropoxyphene | 100.00 | 150.00 | 200.00 | 250.00 | 500.00 | 1000.00 | 2000.00 |
| Oxazepam | 10.00 | 15.00 | 20.00 | 25.00 | 50.00 | 100.00 | 200.00 |
| Oxycodone | 50.00 | 75.00 | 100.00 | 125.00 | 250.00 | 500.00 | 1000.00 |
| Oxymorphone | 50.00 | 75.00 | 100.00 | 125.00 | 250.00 | 500.00 | 1000.00 |
| PCP | 12.50 | 18.75 | 25.00 | 31.25 | 62.50 | 125.00 | 250.00 |
| Propoxyphene | 100.00 | 150.00 | 200.00 | 250.00 | 500.00 | 1000.00 | 2000.00 |
| Sufentanil | 2.50 | 3.75 | 5.00 | 6.25 | 12.50 | 25.00 | 50.00 |
| Temazepam | 10.00 | 15.00 | 20.00 | 25.00 | 50.00 | 100.00 | 200.00 |
| THC-OH | 15.00 | 22.50 | 30.00 | 37.50 | 75.00 | 150.00 | 300.00 |
| THC-COOH | 15.00 | 22.50 | 30.00 | 37.50 | 75.00 | 150.00 | 300.00 |
| Tramadol | 50.00 | 75.00 | 100.00 | 125.00 | 250.00 | 500.00 | 1000.00 |

Note: All concentrations are stated in ng/mL

7. Ab Sciex 5500 QTrap Mass spectrometer with TurboIonSpray (TIS) electrode, positive ionization mode.

8. Shimadzu Prominence HPLC system, consisting of a reagent tray, DGU-20ASR Degasser, CTO-20 AC Column oven, two LC-20ADXR HPLC pumps, and SIL-20ACXR Autosampler.

# 3   Methods

### 3.1   Sample Extraction

1. Label sufficient number of 1.5 mL autosampler injection vials for the double blank (BB), blank (B), seven levels of calibrators, urine negative, QCs, and unknown samples.

2. Remove working calibrators, QCs, and unknown samples from refrigerator. Let them warm to room temperature and vortex the samples for 4–5 s.

3. To the autosampler vial labeled BB, add 1 mL DI water.

4. To the autosampler vial labeled B add 1 mL IS Working solution.

5. To the other vials add 50 μL of calibrators, QCs, and unknowns. To each vial add 1.0 mL of IS Working solution.

6. Cap vials and vortex them for 4–5 s.

7. See **Note 4**.

232     Geza S. Bodor

**3.2 Chromatography (LC) Method Parameters**

1. Pumping Mode: Binary Flow.
2. Total Flow: 0.600 mL/min.
3. Pump B Starting Concentration: 10.0 %.
4. For chromatography buffer gradient *see* Table 5.

**3.3 Autosampler**

Autosampler parameters are listed in Table 6.

**3.4 MS Method Parameters**

MS acquisition is performed using electrospray (Turbo Ion Spray, TIS) ionization in the Positive mode. Data are collected during a single period, single experiment acquisition using scheduled MRM (sMRM).

MS method parameters are listed in Table 7.

**3.5 Source (Compound-Independent) Parameters**

Compound-independent parameters are listed in Table 8.

**3.6 MRM Transitions and Compound-Specific Parameters**

Q1/Q3 masses and compound-specific parameters for analytes are listed in Table 9.

Q1/Q3 masses and compound-specific parameters for internal standards are listed in Table 10. *See* **Notes 5 and 6**.

**3.7 List of Analytes with Their Respective Internal Standards**

The list of analytes with their respective IS are presented in Table 11.

**3.8 Data Analysis**

1. Use Analyst 1.6.2 or Multiquant 3.2 software (or equivalent) for data analysis. The quantifying ions of the analytes (MRM 1 in Table 9) along with the MRM of the corresponding

**Table 5**
**Chromatography buffer gradient**

| Time after injection (min) | Buffer B (%) |
| --- | --- |
| 0.00 | 10 |
| 1.00 | 10 |
| 5.00 | 90 |
| 5.01 | 100 |
| 7.00 | 100 |
| 7.01 | 10 |
| 9.00 | 10 |

**Table 6**
**Autosampler setting**

| Parameter | Value |
|---|---|
| Rinse volume | 2000 μL |
| Needle stroke | 2 mm |
| Rinse speed | 35 μL/s |
| Sampling speed | 15.0 μL/s |
| Purge time | 1.5 min |
| Needle dip time | 15 s |
| Rinse mode | Before and after aspiration |
| Cooler temperature | 4 °C |
| Column oven temp | 40 °C |
| Injection volume | 5.00 μL |

**Table 7**
**MS method parameters**

| MS parameter | Value |
|---|---|
| Scan in period | 1100 |
| MRM detection window | 30 s for each peak |
| Target scan time | 0.300 s |
| Resolution Q1 | Unit |
| Resolution Q3 | Unit |
| MR pause | 5.007 msec |
| MCA | No |
| Detector | 2200 V |

**Table 8**
**Compound-independent source parameters**

| Source parameter | Value |
|---|---|
| Curtain gas (CUR) | 35.00 |
| Collision gas (CAD) | Medium |
| Ion spray voltage (IS) | 4500 V |
| Source temperature (TEM) | 600 °C |
| Ion source gas 1 (GS1) | 55.00 |
| Ion source gas 2 (GS2) | 60.00 |

**Table 9**
**Q1/Q3 masses, RTs, and compound-specific parameters for both MRM transitions of each analyte**

| Analyte name w/MRM # | Q1 Mass | Q3 Mass | RT | DP | EP | CE | CXP |
|---|---|---|---|---|---|---|---|
| 6-MAM 1 | 328.0 | 165.1 | 3.0 | 90 | 10 | 46 | 8 |
| 6-MAM 2 | 328.0 | 211.0 | 3.0 | 90 | 10 | 34 | 8 |
| Alpha Hydroxyalprazolam 1 | 325.2 | 297.1 | 4.9 | 51 | 11 | 37 | 6 |
| Alpha Hydroxyalprazolam 2 | 325.2 | 216.1 | 4.9 | 51 | 11 | 53 | 4 |
| Alprazolam 1 | 309.1 | 205.1 | 5.0 | 50 | 10 | 56 | 15 |
| Alprazolam 2 | 309.1 | 281.0 | 5.0 | 50 | 10 | 38 | 15 |
| Amphetamine 1 | 136.1 | 91.0 | 2.8 | 121 | 10 | 21 | 12 |
| Amphetamine 2 | 136.1 | 119.1 | 2.8 | 121 | 10 | 11 | 6 |
| Benzoylecgonine 1 | 290.1 | 168.1 | 3.6 | 200 | 10 | 25 | 15 |
| Benzoylecgonine 2 | 290.1 | 105.1 | 3.6 | 150 | 10 | 40 | 8 |
| Buprenorphine 1 | 468.1 | 396.1 | 4.5 | 60 | 10 | 54 | 15 |
| Buprenorphine 2 | 468.1 | 414.0 | 4.5 | 60 | 10 | 45 | 15 |
| Carisoprodol 1 | 261.1 | 176.1 | 4.6 | 300 | 15 | 12 | 15 |
| Carisoprodol 2 | 261.1 | 97.0 | 4.6 | 100 | 10 | 21 | 15 |
| Clonazepam 1 | 316.0 | 269.9 | 4.9 | 50 | 10 | 35 | 15 |
| Clonazepam 2 | 316.0 | 214.1 | 4.9 | 50 | 10 | 50 | 15 |
| Codeine 1 | 300.0 | 152.1 | 2.8 | 90 | 10 | 80 | 15 |
| Codeine 2 | 300.0 | 115.1 | 2.8 | 90 | 10 | 85 | 15 |
| Diazepam 1 | 285.1 | 193.1 | 5.3 | 50 | 10 | 45 | 15 |
| Diazepam 2 | 285.1 | 154.1 | 5.3 | 50 | 10 | 35 | 15 |
| EDDP 1 | 278.1 | 234.1 | 4.5 | 300 | 5 | 43 | 15 |
| EDDP 2 | 278.1 | 186.1 | 4.5 | 250 | 10 | 45 | 15 |
| Fentanyl 1 | 337.1 | 188.1 | 4.3 | 41 | 10 | 31 | 8 |
| Fentanyl 2 | 337.1 | 105.2 | 4.3 | 41 | 10 | 47 | 10 |
| Flunitrazepam 1 | 314.0 | 268.1 | 5.0 | 50 | 10 | 36 | 15 |
| Flunitrazepam 2 | 314.0 | 239.0 | 5.0 | 50 | 10 | 45 | 15 |
| Flurazepam 1 | 388.0 | 315.1 | 4.6 | 200 | 10 | 32 | 15 |
| Flurazepam 2 | 388.0 | 134.2 | 4.6 | 50 | 10 | 65 | 15 |
| Hydrocodone 1 | 300.0 | 199.1 | 3.1 | 72 | 10 | 54 | 12 |
| Hydrocodone 2 | 300.0 | 128.1 | 3.1 | 60 | 10 | 75 | 12 |

(continued)

**Table 9**
**(continued)**

| Analyte name w/MRM # | Q1 Mass | Q3 Mass | RT | DP | EP | CE | CXP |
|---|---|---|---|---|---|---|---|
| Hydromorphone 1 | 286.1 | 185.1 | 2.1 | 91 | 10 | 39 | 12 |
| Hydromorphone 2 | 286.1 | 128.1 | 2.1 | 91 | 10 | 75 | 10 |
| Lorazepam 1 | 321.1 | 275.1 | 4.8 | 100 | 10 | 30 | 20 |
| Lorazepam 2 | 321.1 | 229.1 | 4.8 | 100 | 10 | 41 | 20 |
| MDA 1 | 180.1 | 105.1 | 3.1 | 180 | 10 | 29 | 8 |
| MDA 2 | 180.1 | 133.1 | 3.1 | 200 | 10 | 23 | 10 |
| MDEA 1 | 208.0 | 163.1 | 3.5 | 200 | 10 | 19 | 12 |
| MDEA 2 | 208.0 | 105.1 | 3.5 | 150 | 10 | 35 | 8 |
| MDMA 1 | 193.9 | 105.0 | 3.3 | 55 | 6 | 30 | 15 |
| MDMA 2 | 193.9 | 135.0 | 3.3 | 80 | 11 | 30 | 20 |
| Meperidine 1 | 248.1 | 174.1 | 4.0 | 100 | 10 | 47 | 20 |
| Meperidine 2 | 248.1 | 220.1 | 4.0 | 72 | 10 | 49 | 8 |
| Meprobamate 1 | 219.1 | 97.0 | 4.0 | 200 | 10 | 19 | 15 |
| Meprobamate 2 | 219.1 | 158.1 | 4.0 | 150 | 10 | 12 | 15 |
| Methadone 1 | 310.2 | 105.1 | 4.9 | 200 | 10 | 32 | 15 |
| Methadone 2 | 310.2 | 265.1 | 4.9 | 200 | 10 | 21 | 15 |
| Methamphetamine 1 | 150.2 | 64.9 | 3.1 | 60 | 9 | 55 | 20 |
| Methamphetamine 2 | 150.2 | 91.0 | 3.1 | 130 | 9 | 25 | 35 |
| Midazolam 1 | 326.0 | 291.1 | 4.8 | 101 | 10 | 37 | 22 |
| Midazolam 2 | 326.0 | 249.1 | 4.8 | 101 | 10 | 49 | 18 |
| Morphine 1 | 286.0 | 152.0 | 1.7 | 91 | 10 | 75 | 12 |
| Morphine 2 | 286.0 | 165.0 | 1.7 | 91 | 10 | 49 | 12 |
| Naloxone 1 | 328.1 | 212.1 | 2.7 | 65 | 10 | 50 | 15 |
| Naloxone 2 | 328.1 | 253.1 | 2.7 | 65 | 10 | 32 | 15 |
| Naltrexone 1 | 342.1 | 267.2 | 3.0 | 86 | 10 | 39 | 18 |
| Naltrexone 2 | 342.1 | 282.1 | 3.0 | 86 | 10 | 37 | 20 |
| Norbuprenorphine 1 | 414.1 | 186.9 | 4.2 | 50 | 10 | 50 | 20 |
| Norbuprenorphine 2 | 414.1 | 381.9 | 4.2 | 40 | 7 | 35 | 20 |
| Nordiazepam 1 | 271.0 | 140.0 | 5.1 | 110 | 13 | 35 | 13 |
| Nordiazepam 2 | 271.0 | 165.1 | 5.1 | 100 | 7.5 | 35 | 20 |
| Norfentanyl 1 | 233.0 | 84.2 | 3.6 | 55 | 10 | 21 | 15 |

(continued)

**Table 9**
**(continued)**

| Analyte name w/MRM # | Q1 Mass | Q3 Mass | RT | DP | EP | CE | CXP |
|---|---|---|---|---|---|---|---|
| Norfentanyl 2 | 233.0 | 150.1 | 3.6 | 55 | 10 | 28 | 15 |
| Normeperidine 1 | 234.0 | 160.0 | 3.9 | 200 | 10 | 23 | 14 |
| Normeperidine 2 | 234.0 | 188.1 | 3.9 | 66 | 10 | 21 | 16 |
| Norpropoxyphene 1 | 308.0 | 100.1 | 4.6 | 120 | 7 | 20 | 20 |
| Norpropoxyphene 2 | 308.0 | 143.0 | 4.6 | 110 | 7 | 30 | 20 |
| Oxazepam 1 | 286.9 | 241.1 | 5.0 | 55 | 10 | 30 | 11 |
| Oxazepam 2 | 286.9 | 269.2 | 5.0 | 55 | 10 | 19 | 9 |
| Oxycodone 1 | 316.1 | 241.1 | 3.0 | 50 | 10 | 38 | 11 |
| Oxycodone 2 | 316.1 | 256.1 | 3.0 | 50 | 10 | 35 | 11 |
| Oxymorphone 1 | 302.1 | 227.1 | 1.9 | 75 | 10 | 39 | 8 |
| Oxymorphone 2 | 302.1 | 198.0 | 1.9 | 75 | 10 | 57 | 8 |
| PCP 1 | 244.1 | 91.0 | 4.3 | 45 | 10 | 18 | 15 |
| PCP 2 | 244.1 | 159.1 | 4.3 | 100 | 10 | 19 | 15 |
| Propoxyphene 1 | 340.1 | 266.2 | 4.7 | 200 | 10 | 10 | 15 |
| Propoxyphene 2 | 340.1 | 91.1 | 4.7 | 44 | 10 | 70 | 15 |
| Sufentanil 1 | 387.1 | 238.0 | 4.7 | 46 | 10 | 27 | 16 |
| Sufentanil 2 | 387.1 | 111.1 | 4.7 | 46 | 10 | 49 | 10 |
| Temazepam 1 | 301.1 | 177.1 | 5.1 | 70 | 10 | 51 | 20 |
| Temazepam 2 | 301.1 | 255.1 | 5.1 | 200 | 10 | 30 | 8 |
| THC-COOH 1 | 345.2 | 193.2 | 5.6 | 54 | 10 | 33 | 10 |
| THC-COOH 2 | 345.2 | 299.2 | 5.6 | 54 | 10 | 33 | 10 |
| THC-OH 1 | 331.2 | 193.2 | 5.6 | 70 | 11 | 30 | 15 |
| THC-OH 2 | 331.2 | 313.0 | 5.6 | 70 | 8 | 18 | 20 |
| Tramadol 1 | 264.1 | 58.1 | 3.8 | 45 | 10 | 115 | 10 |
| Tramadol 2 | 264.1 | 42.1 | 3.8 | 45 | 10 | 110 | 10 |

**Note**: Numbers 1 and 2 indicate MRM transitions 1 and 2, respectively

deuterated IS are used to construct the 7-point calibration curves by calculating the calibrator/IS peak area ratios against the nominal concentrations of the calibrators. The unknown concentrations are read from this seven-point calibration curve.

2. Typical calibration curve correlation coefficients are >0.99.

**Table 10**
**Q1/Q3 masses, RTs, and compound-specific parameters of the internal standards**

| IS name w/MRM | Q1 Mass | Q3 Mass | RT | DP | EP | CE | CXP |
|---|---|---|---|---|---|---|---|
| Amphetamine-d5 | 141.1 | 93.0 | 2.7 | 126 | 10 | 23 | 12 |
| Benzoylecgonine-D8 | 298.1 | 171.1 | 3.6 | 200 | 10 | 25 | 15 |
| Codeine-D6 | 306.2 | 152.2 | 2.8 | 90 | 10 | 80 | 15 |
| Fentanyl-D5 | 342.2 | 105.1 | 4.3 | 50 | 10 | 52 | 15 |
| Hydrocodone-D6 | 306.0 | 199.1 | 3.1 | 86 | 10 | 39 | 8 |
| Hydromorphone-D6 | 292.0 | 185.0 | 2.1 | 100 | 10 | 39 | 12 |
| Meperidine-D4 | 252.2 | 224.1 | 3.9 | 250 | 10 | 29 | 16 |
| Methadone-D3 | 313.2 | 105.0 | 4.8 | 250 | 10 | 38 | 20 |
| Methamphetamine-D5 | 155.1 | 121.2 | 3.1 | 80 | 8 | 17 | 30 |
| Morphine-D6 | 292.1 | 152.0 | 1.7 | 65 | 10 | 75 | 15 |
| Nordiazepam-D5 | 276.1 | 140.1 | 5.1 | 56 | 8 | 37 | 4 |
| Norpropoxyphene-D5 | 313.2 | 100.1 | 4.6 | 250 | 10 | 17 | 15 |
| PCP-D5 | 249.1 | 96.0 | 4.3 | 30 | 10 | 55 | 16 |
| THC-COOH-D3 | 348.2 | 196.1 | 5.6 | 62 | 10 | 35 | 15 |
| THC-OH-D3 | 334.2 | 196.3 | 5.6 | 55 | 10 | 36 | 15 |
| Tramadol-13C, D3 | 268.3 | 58.1 | 3.8 | 100 | 10 | 43 | 20 |
| Norbuprenorphine-D3 | 417.2 | 364.1 | 4.2 | 180 | 10 | 40 | 20 |

3. Typical intra- and inter-assay imprecision are <10 % but many analytes will have better than 5 % total imprecision.

4. Quality control ranges are established by establishing the mean and standard deviation for each analyte from at least 20 measurements. Acceptable range is mean ± 2SD.

# 4 Notes

1. This chromatographic separation method is very sensitive to the composition of the mobile phase. A small change in buffer composition, especially that of Buffer B, can cause significant retention time shift of some analytes, causing some peaks to partially shift outside of the established RT window (i.e., Midazolam) and/or the integration of the wrong peak (i.e., Codeine could be mis-integrated for Hydrocodone). Buffer composition change is greatly accelerated by elevated ambient temperatures. Careful review of peaks before releasing results is mandatory to detect RT shift.

2. The Calibrator Stock (100×) solution may be used to perform carry-over studies if sufficiently high concentration patient sample is not available for certain drug(s).

3. Calibrator C can be considered as the "cut-off concentration" if this method is used for screening. The High Calibrator

**Table 11**
**List of analytes with their respective IS**

| Internal standard (IS) | Analyte |
|---|---|
| 6-MAM-D3 | 6-MAM |
|  | Naltrexone |
| Amphetamine-D5 | Amphetamine |
| Benzoylecgonine-D8 | Benzoylecgonine |
|  | Meprobamate |
| Codeine-D6 | Codeine |
|  | Naloxone |
|  | Oxycodone |
| Fentanyl-D5 | Carisoprodol |
|  | Fentanyl |
|  | Norfentanyl |
|  | Sufentanil |
| Hydrocodone-D6 | Hydrocodone |
| Hydromorphone-D6 | Hydromorphone |
| Meperidine-D4 | Meperidine |
|  | Normeperidine |
| Methadone-D3 | EDDP |
|  | Methadone |
| Methamphetamine-D5 | MDA |
|  | MDEA |
|  | MDMA |
|  | Methamphetamine |
| Morphine-D6 | Morphine |
|  | Oxymorphone |
| Norbuprenorphine-D3 | Buprenorphine |
|  | Norbuprenorphine |

(continued)

**Table 11**
**(continued)**

| Internal standard (IS) | Analyte |
|---|---|
| Nordiazepam-D5 | Alpha Hydroxyalprazolam |
| | Alprazolam |
| | Clonazepam |
| | Diazepam |
| | Flunitrazepam |
| | Flurazepam |
| | Lorazepam |
| | Midazolam |
| | Nordiazepam |
| | Oxazepam |
| | Temazepam |
| Norpropoxyphene-D5 | Norpropoxyphene |
| | Propoxyphene |
| PCP-D5 | PCP |
| THC-COOH-D3 | THC-COOH |
| THC-OH-D3 | THC-OH |
| Tramadol-13C, D3 | Tramadol |

(Calibrator G) solution contains 10× the cutoff concentration (Calibrator C) of each drug in the panel.

4. Extracted samples should be injected as soon as possible following extraction. However, extracted samples may be delayed for instrument maintenance or other reasons. We evaluated stability of extracted samples and found that the samples are stable for at least 7 days in the unopened autosampler vials in the refrigerator or on the refrigerated autosampler.

5. All compound-specific method parameters were developed and optimized in the Analyst 1.6.2 software and are stored in the appropriate acquisition and quantitation method files and are specific for the AB Sciex QTrap 5500 instrument. Because a large number of analytes are measured simultaneously and they cover a wide dynamic range, mass spectrometric parameters were optimized for each analyte to provide the best response ratio throughout the clinically important concentration ranges. This optimization required tuning down instrument sensitivity

for some drugs to prevent detector saturation at high analyte concentrations. Additional optimization by the user of this method may be required if different instrumentation is used than what is described in this method.

6. The listed retention times (RT) are observed on our multiplexed, dual Shimadzu HPLC system. Because of the extended tubing of a multiplexed system these retention times may not be applicable to a single HPLC system. If adapting this method to a non-multiplexed system, the user must establish the appropriate retention times for their hardware.

## References

1. Institute of Medicine Report from the Committee on Advancing Pain Research, Care, and Education: Relieving pain in America, a blueprint for transforming prevention, care, education and research. The National Academies Press, 2011. http://books.nap.edu/openbook.php?record_id=13172&page=1

2. National Centers for health statistics, chartbook on trends in the health of Americans 2006, special feature: pain. http://www.cdc.gov/nchs/data/hus/hus06.pdf

3. The management of chronic pain in patients with breast cancer. The Steering Committee on Clinical Practice Guidelines for the Care and Treatment of Breast Cancer. Canadian Society of Palliative Care Physicians. Canadian Association of Radiation Oncologists. CMAJ 1998;158 Suppl 3:S71-81

4. Peppin JF et al Recommendations for urine drug monitoring as a component of opioid therapy in the treatment of chronic pain. Pain Med 2012;13:886–896

# Chapter 25

# Quantification of Free Phenytoin by Liquid Chromatography Tandem Mass Spectrometry (LC/MS/MS)

## Judy Peat, Clint Frazee, and Uttam Garg

## Abstract

Phenytoin (diphenylhydantoin) is an anticonvulsant drug that has been used for decades for the treatment of many types of seizures. The drug is highly protein bound and measurement of free-active form of the drug is warranted particularly in patients with conditions that can affect drug protein binding. Here, we describe a LC/MS/MS method for the measurement of free phenytoin. Free drug is separated by ultrafiltration of serum or plasma. Ultrafiltrate is treated with acetonitrile containing internal standard phenytoin d-10 to precipitate proteins. The mixture is centrifuged and supernatant is injected onto LC-MS-MS, and analyzed using multiple reaction monitoring. This method is linear from 0.1 to 4.0 μg/mL and does not demonstrate any significant ion suppression or enhancement.

**Key words** Phenytoin, Free phenytoin, Mass spectrometry, Seizures, Convulsions, Epilepsy

## 1 Introduction

Phenytoin, 5,5-diphenylimidazolidinedione, is an anticonvulsant drug frequently prescribed for grand mal epilepsy, cortical focal seizures, and temporal lobe epilepsy [1, 2]. It was first synthesized in 1908 and continues to be a preferred drug in the management of many types of epilepsy. Since the drug has narrow therapeutic window, its therapeutic drug monitoring is desired [1]. Generally, total phenytoin concentration is measured and is adequate for therapeutic drug monitoring. However, since the drug is highly protein bound (>90 %), and many conditions such as altered protein concentrations, renal dysfunction, and co-administration of drugs that bind to albumin can affect free phenytoin concentration, measurement of free (active) drug is required for optimal therapeutic drug monitoring [3, 4]. The optimal therapeutic concentration for total and free phenytoin is 10–20 μg/mL and 1–2 μg/mL respectively [1, 2].

Phenytoin is generally measured by immunoassays or chromatographic methods [5–12]. Immunoassays for total phenytoin are

Uttam Garg (ed.), *Clinical Applications of Mass Spectrometry in Drug Analysis: Methods and Protocols*, Methods in Molecular Biology, vol. 1383, DOI 10.1007/978-1-4939-3252-8_25, © Springer Science+Business Media New York 2016

readily available, whereas immunoassays for free phenytoin are scanty on automated chemistry analyzers. Also, immunoassays are not very specific and may exhibit interferences [13, 14]. Therefore, chromatographic methods are preferred over immunoassays. Various chromatographic methods such as gas-chromatography mass spectrometry, HPLC with UV detector or LC/MS/MS methods have been described. We present a simple LC/MS/MS method for the assay of free phenytoin.

## 2   Materials

### 2.1   Samples

0.5 mL heparinized plasma or serum in no-gel plain tube. Store in a refrigerator (<5 °C) until analysis. The samples are stable for 4 days.

### 2.2   Solvents and Reagents

1. Sodium phosphate-anhydrous, dibasic ($Na_2HPO_4$) (ACS grade).

2. Sodium phosphate-monohydrate, monobasic ($NaH_2PO_4 \bullet H_2O$) (ACS grade).

3. Sodium azide (Mallincroft).

4. Human drug-free pooled normal plasma (UTAK).

5. 0.20 M $Na_2HPO_4$: Add 2.84 g of $Na_2HPO_4$ (anhydrous) to a 100 mL volumetric flask and qs with DI water. Stopper and invert gently until mixed.

6. 0.20 M $NaH_2PO_4$: Add 1.38 g $NaH_2PO_4 \bullet H_2O$ (monohydrate) to a 50 mL volumetric flask and qs with DI water. Stopper and invert gently until mixed.

7. 0.20 M Phosphate Buffer: Combine 2 mL of 0.20 M $NaH_2PO_4$ and 100 mL of 0.20 M $Na_2HPO_4$ in a 150 mL beaker. Mix and check pH. The pH should be 7.4 ± 0.1. Adjust pH to 7.4 with 0.20 M $Na_2HPO_4$ or 0.20 M $NaH_2PO_4$, if necessary.

8. 0.6 M NaCl: Add 3.6 g of NaCl to a 100 mL volumetric flask and qs with DI water. Stopper and invert gently until mixed.

9. Phosphate buffered saline, pH 7.4 (used for making calibrators): Combine 100 mL of phosphate buffer with 100 mL of 0.6 M NaCl. Add 50 mg of sodium azide. Mix and check pH. The pH should be 7.4 ± 0.1. Adjust pH to 7.4 with 0.20 M $Na_2HPO_4$ or 0.20 M $NaH_2PO_4$, if necessary.

10. Mobile phase A: Add 1.14 mL formic acid to 1 L of HPLC grade water. Mix, filter, and degas. Use at room temperature. Stable for 6 months.

11. Mobile phase B: Add 1.14 mL formic acid to 1 L of HPLC grade methanol. Mix, filter, and degas. Use at room temperature. Stable for 6 months.

**Table 1**
**Preparation of calibrators in phosphate buffered saline**

| Calibrator (µg/ mL) | mL of Phosphate buffer, pH 7.4 | µL of 2⁰ Standard | µL of 3⁰ Standard |
|---|---|---|---|
| Blank | 1.00 | | |
| 0.1 | 0.99 | | 10 |
| 0.2 | 0.98 | | 20 |
| 0.5 | 0.95 | | 50 |
| 1.0 | 0.90 | | 100 |
| 2.0 | 0.98 | 20 | |
| 4.0 | 0.96 | 40 | |

Note: Calibrators are stable for 6 months when stored at < 20 °C

**2.3 Standards and Internal Standards**

1. Primary Internal Standard: 100 µg/mL phenytoin-d10 in methanol (Cerilliant). For stability see Certificate of Analysis.

2. Primary Standard: 1 mg/mL phenytoin in methanol (Cerilliant). For stability see Certificate of Analysis.

3. Secondary (2⁰): 100 µg/mL phenytoin in methanol. Make a 1:10 quantitative dilution of the primary phenytoin standard. Stable for 1 year when stored at < 20 °C.

4. Tertiary Standard (3⁰): 10 µg/mL phenytoin in methanol. Make a 1:10 quantitative dilution of the secondary phenytoin standard. Stable for 1 year when stored at < 20 °C.

**2.4 Calibrators and Controls**

1. Calibrators: Prepare calibrators 1–6 according to Table 1.

2. Quality Controls: BIORAD Liquichek TDM Levels 1 and 2. Ranges are established in-house. Store at < 0 °C.

**2.5 Analytical Equipment and Supplies**

1. Applied Biosystems LC/MS/MS 4000Q TRAP with Shimadzu HPLC.

2. Analytical column: Restek Ultra BiPh 5 µm 50 × 2, 1 mm.

3. 1.5 mL microcentrifuge tubes.

4. Autosampler vials with glass inserts and screw caps.

# 3 Methods

**3.1 Stepwise Procedure**

1. Add 0.5 mL of sample, blank, and controls to labeled Millipore Protein Filter.

2. Centrifuge for 10 min at 2000 × $g$ at 20 °C.

**Table 2**
**LC parameters**

| Time (min) | Mobile phase A (%) | Mobile phase B (%) |
|------------|--------------------|--------------------|
| 0.2        | 98                 | 2                  |
| 1.0        | 5                  | 95                 |
| 2.0        | 5                  | 95                 |
| 3.5        | 98                 | 2                  |
| 3.6        | Stop               | Stop               |

Total flow: 0.6 mL/min
Oven: 40 °C

3. Pipette 20 μL of each patient, blank, and control ultrafiltrate, and 20 μL of each buffered calibrator (unfiltered), into labeled microcentrifuge tubes.

4. Pipette 100 μL of IS reagent into each microcentrifuge tube.

5. Vortex and let stand for 5 min. Centrifuge for 5 min at $10,000 \times g$ (*See* **Note 1**).

6. Pipette approximately 80 μL of supernatant into a labeled autosampler vial.

7. Inject 5 μL into LC/MS/MS for analysis (LC parameters are given in Table 2).

*3.2  Data Analysis*

1. Data are analyzed using Target Software (Thru-Put Systems) or similar software.

2. Standard curves are generated based on linear regression of the analyte/IS peak area ratio (y) versus analyte concentration (x) using the quantifying ion listed in Table 3. Monitored ions are given in Table 3. *See* **Notes 2** and **3.**

3. Typical ion extract chromatogram is shown in Fig. 1.

4. The coefficient of correlation must be >0.99.

5. Linearity of the method is 0.1–4.0 μg/mL.

6. Runs are accepted if calculated controls fall within two standard deviations of target values (*See* **Note 4**).

# 4  Notes

1. There should be no precipitate present in the microcentrifuge tube after centrifugation. If precipitation is observed, the Millipore Protein Filter may have been faulty. A fresh sample aliquot should be prepared using a new filter.

**Table 3**
**Mass spectrometer parameters**

| Q1 mass | Q3 mass | Dwell (msec) | Parameter | Start | Stop | ID |
|---------|---------|--------------|-----------|-------|------|-----|
| 252.98 | 182.10 | 200 | CE\<p\>CXP | 25.00\<p\>10.00 | 25.00\<p\>10.00 | Phenytoin Quant ion |
| 252.98 | 104.00 | 200 | CE\<p\>CXP | 47.00\<p\>16.00 | 47.00\<p\>16.00 | Phenytoin Qual Ion |
| 263.10 | 192.16 | 200 | CE\<p\>CXP | 27.00\<p\>16.00 | 27.00\<p\>16.00 | Phenytoin-d10 |

Scan type: MRM
Polarity: Positive
Ion Source: Turbo Spray
Resolution Q1 and Q3: Unit
CUR: 25.00
TEM: 550.00
GS1: 50.00
GS2: 50.00
ihe: ON
CAD: Medium
IS: 4000.00
DP: 71.00
EP: 10.00

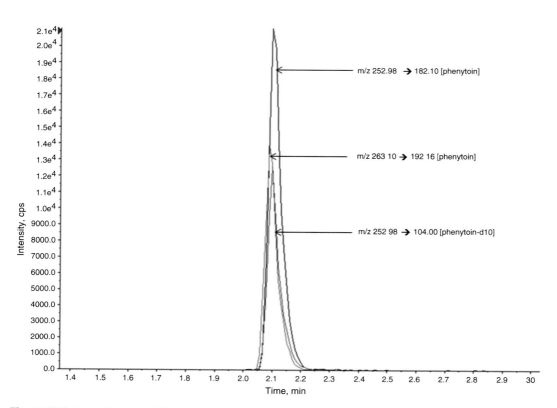

**Fig. 1** MRM chromatogram of phenytoin and phenytoin-d10

2. A value for total phenytoin can be obtained by analyzing 20 μL of unfiltered sample. If a dilution is needed blank plasma may be used.

3. Water can be used as a diluent for the ultrafiltrate when a free phenytoin value is above the curve.

4. Deviation is generally <10 %.

## References

1. Broussard LA (2007) Monitoring anticonvulsants concentrations: general considerations. In: Hammett-Stabler CA, Dasgupta A (eds) Therapeutic drug monitoring data. AACC Press, Washington, D.C., pp 41–75

2. Moyer TP, Shaw LM (2006) Therapeutic drugs and their management. In: Burtis CA, Ashwood ER, Bruns DE (eds) Tietz textbook of clinical chemistry and molecular diagnostics, 4th edn. Elsevier Saunders, St. Louis, MO, pp 1237–1285

3. Biddle DA, Wells A, Dasgupta A (2000) Unexpected suppression of free phenytoin concentration by salicylate in uremic sera due to the presence of inhibitors: MALDI mass spectrometric determination of molecular weight range of inhibitors. Life Sci 66:143–151

4. Wolf GK, McClain CD, Zurakowski D, Dodson B, McManus ML (2006) Total phenytoin concentrations do not accurately predict free phenytoin concentrations in critically ill children. Pediatr Crit Care Med 7:434–439, quiz 440

5. Bardin S, Ottinger JC, Breau AP, O'Shea TJ (2000) Determination of free levels of phenytoin in human plasma by liquid chromatography/tandem mass spectrometry. J Pharm Biomed Anal 23:573–579

6. Godolphin W, Trepanier J (1984) Gas chromatography versus immunoassay (EMIT) for analysis of free phenytoin in serum ultrafiltrate. Ther Drug Monit 6:374–375

7. Joern WA (1981) Gas-chromatographic assay of free phenytoin in ultrafiltrates of plasma: test of a new filtration apparatus and specimen stability. Clin Chem 27:417–421

8. Lin PC, Hsieh YH, Liao FF, Chen SH (2010) Determination of free and total levels of phenytoin in human plasma from patients with epilepsy by MEKC: an adequate alternative to HPLC. Electrophoresis 31:1572–1582

9. May T, Rambeck B (1990) Fluctuations of unbound and total phenytoin concentrations during the day in epileptic patients on valproic acid comedication. Ther Drug Monit 12:124–128

10. Van Langenhove A, Costello CE, Biller JE, Biemann K, Browne TR (1980) A mass spectrometric method for the determination of stable isotope labeled phenytoin suitable for pulse dosing studies. Biomed Mass Spectrom 7:576–581

11. Van Langenhove A, Costello CE, Biller JE, Biemann K, Browne TR (1981) A gas chromatographic/mass spectrometric method for the simultaneous quantitation of 5,5-diphenylhydantoin (phenytoin), its para-hydroxylated metabolite and their stable isotope labelled analogs. Clin Chim Acta 115:263–275

12. Garg U, Peat J, Frazee C 3rd, Nguyen T, Ferguson AM (2013) A simple isotope dilution electrospray ionization tandem mass spectrometry method for the determination of free phenytoin. Ther Drug Monit 35:831–835

13. Datta P (1997) Oxaprozin and 5-(p-hydroxyphenyl)-5-phenylhydantoin interference in phenytoin immunoassays. Clin Chem 43:1468–1469

14. Datta P, Scurlock D, Dasgupta A (2005) Analytic performance evaluation of a new turbidimetric immunoassay for phenytoin on the ADVIA 1650 analyzer: effect of phenytoin metabolite and analogue. Ther Drug Monit 27:305–308

# Chapter 26

# Detection of Stimulants and Narcotics by Liquid Chromatography-Tandem Mass Spectrometry and Gas Chromatography-Mass Spectrometry for Sports Doping Control

## Brian D. Ahrens, Yulia Kucherova, and Anthony W. Butch

## Abstract

Sports drug testing laboratories are required to detect several classes of compounds that are prohibited at all times, which include anabolic agents, peptide hormones, growth factors, beta-2 agonists, hormones and metabolic modulators, and diuretics/masking agents. Other classes of compounds such as stimulants, narcotics, cannabinoids, and glucocorticoids are also prohibited, but only when an athlete is in competition. A single class of compounds can contain a large number of prohibited substances and all of the compounds should be detected by the testing procedure. Since there are almost 70 stimulants on the prohibited list it can be a challenge to develop a single screening method that will optimally detect all the compounds. We describe a combined liquid chromatography-tandem mass spectrometry (LC-MS/MS) and gas chromatography-mass spectrometry (GC-MS) testing method for detection of all the stimulants and narcotics on the World Anti-Doping Agency prohibited list. Urine for LC-MS/MS testing does not require sample pretreatment and is a direct dilute and shoot method. Urine samples for the GC-MS method require a liquid-liquid extraction followed by derivatization with trifluoroacetic anhydride.

**Key words** Liquid chromatography, Gas chromatography, Mass spectrometry, Stimulants, Narcotics, Urine, Doping, World Anti-Doping Agency

## 1 Introduction

World Anti-Doping Agency (WADA)-accredited drug testing laboratories are required to monitor numerous classes of compounds for sports doping control programs based on the WADA prohibited list of substances (Table 1) [1]. For athletes that are in competition, additional classes of compounds are also prohibited such as stimulants, narcotics, cannabinoids, glucocorticoids, and beta-blockers (Table 1). Some of the classes of compounds on the prohibited list contain large numbers of drugs that need to be monitored for doping control programs. For example, there are close to 70 stimulants on the WADA prohibited list and the number

Uttam Garg (ed.), *Clinical Applications of Mass Spectrometry in Drug Analysis: Methods and Protocols*, Methods in Molecular Biology, vol. 1383, DOI 10.1007/978-1-4939-3252-8_26, © Springer Science+Business Media New York 2016

**Table 1**
**2015 World Anti-doping Agency list of prohibited substances**

| *Prohibited at all times* |
| --- |
| • Anabolic agents |
| • Peptide hormones, growth factors, related substances and mimetics |
| • Beta-2 agonists |
| • Hormone and metabolic modulators |
| • Diuretics and masking agents |
| • Beta-blockers (archery and shooting only) |
| *Prohibited only in competition* |
| • Stimulants |
| • Narcotics |
| • Cannabinoids |
| • Glucocorticoids |
| • Beta-blockers (some sports) |

of target compounds being monitored by accredited laboratories is even larger since compounds with a "similar chemical structure or biological effect" are also banned by WADA [1].

Given the large number of stimulants that are monitored by anti-doping laboratories it can be a challenge to develop a single screening method to detect all the relevant compounds, especially at the minimum required performance levels required by WADA [2]. Gas chromatography mass spectrometry (GC-MS) is often utilized to detect a wide range of stimulants after solid-phase or liquid-liquid extraction of urine samples [3, 4]. When using GC-MS methods to detect stimulants a derivatized urine extract is usually prepared because some underivatized stimulants produce a limited number of diagnostic ions that are often in low abundance [5]. Liquid chromatography-tandem mass spectrometry (LC-MS/MS) can be used to detect thermolabile, volatile, and polar compounds and does not require sample derivatization for good chromatographic separation and abundant diagnostic ions. LC-MS/MS has routinely been used for detection of stimulants and a major advantage of this approach is that high-throughput screening methods can be developed without the need for a sample concentration/cleanup step (dilute and shoot methods) [6]. Since some compounds such as the metabolites of cocaine and methylphenidate are poorly recovered after liquid-liquid extraction, dilute and shoot LC-MS/MS methods are an attractive alternative for detecting these stimulants at the required WADA minimum required performance levels [7]. Narcotics can also be detected by LC-MS/ MS dilute and shoot methods, which have the necessary analytical sensitivity to detect fentanyl and fentanyl derivatives at the required detection limit of 2 ng/mL [6].

In this chapter we describe a combined LC-MS/MS and GC-MS method for detection of all the stimulants and narcotics on the WADA prohibited list. The LC-MS/MS method does not require a urine pretreatment step (dilute and shoot method). For GC-MS analysis, urine samples require a liquid-liquid extraction step followed by derivatization with trifluoroacetic anhydride.

## 2   Materials

### 2.1   Samples

Only urine samples can be tested using this method. Refrigerated storage of urine is recommended prior to analysis to prevent bacterial growth and urine degradation.

### 2.2   Reagents and Solutions

1. Phenazine working solution (0.5 µg/mL): To a 100 mL volumetric flask add 1 mL of a 0.05 mg/mL phenazine solution prepared in isopropanol, 9 mL of isopropanol, and 90 mL of diethylether.

2. Hydrochloric/mercaptoacetic acid (9:1): To a 1 L conical flask add 275 mL of deionized water and a magnetic stir bar. While stirring slowly, add 25 mL of 12 M hydrochloric acid (HCl) and 32 mL of mercaptoacetic acid.

3. Sodium hydroxide (12 M): To a 1 L conical flask placed in an ice bath add 500 mL of deionized water and a magnetic stir bar. While stirring, add 240 g of sodium hydroxide in 10 g portions and continue stirring until all solids dissolve.

4. Sodium bicarbonate:potassium bicarbonate solid buffer (3:2, w:w).

5. Ammonium bicarbonate buffer (100 mM): To a 1 L volumetric flask add 850 mL of deionized water, a magnetic stir bar and 7.906 g of ammonium bicarbonate. Stir until all solids dissolve and adjust to pH 7.0 by dropwise addition of glacial acetic acid. Bring to 1 L final volume with deionized water, stopper, and invert several times to mix thoroughly.

6. Enzyme:  Purified  aqueous  β-glucuronidase  solution (IMCSzyme product 04-E1F, IMCS) with a specific activity >50,000 U/mL.

7. LC-MS/MS reagent mixture: To a 10 mL glass tube with a screw cap add 50 µL of internal standard solution for LC-MS/MS, 1 mL of 100 mM ammonium bicarbonate buffer and 50 µL of enzyme. Cap the glass tube and invert several times to mix.

8. 0.1 % formic acid in water: Add 999 mL of deionized water to a 1 L glass reagent bottle followed by the addition of 1 mL of formic acid. Mix thoroughly.

9. 0.1 % formic acid in methanol: Add 999 mL of methanol to a 1 L glass reagent bottle followed by the addition of 1 mL of formic acid. Mix thoroughly.

10. Trifluoroacetic anhydride: Reagent Plus ≥99 % from Sigma-Aldrich.

**2.3  Calibrators, Controls, and Internal Standards**

1. Negative control urine: Urine from a healthy volunteer not taking any medications or supplements.

2. GC-MS calibrator for stimulants: Add 2.5 µg/mL of cathine and 5 µg/mL ephedrine, methephedrine and pseudoephedrine, and 3 µg/mL of caffeine to the negative control urine. All drugs are added from solutions prepared in methanol.

3. LC-MS/MS positive control for stimulants: Add 400 ng/mL of dobutamine and 200 ng/mL of amphetamine, benzoylecgonine, chlorphentermine, cropropamide, crotethamide, cyclazodone, dimethylamphetamine, epinephrine, fenbutrazate, fencamine, heptaminol, isometheptene, methylhexanamine, methylphenidate, norphenylephrine, octopamine, pentetrazol, phenpromethazine, ritalinic acid, strychnine, and tuaminoheptane to the negative control urine. All drugs are added from solutions prepared in methanol.

4. LC-MS/MS positive control for fentanyls: Add 2 ng/mL of alfentanil, fentanyl, norfentanyl, remifentanil, and sufentanil to the negative control urine (*see* **Note 1**). All drugs are added from solutions prepared in methanol.

5. LC-MS/MS positive control for narcotics: Add 100 ng/mL of morphine, 50 ng/mL of hydrocodone, hydromorphone, pethidine, noroxycodone, oxycodone, oxymorphone, and propoxyphene, and 5 ng/mL of buprenorphine and norbuprenorphine to the negative control urine. All drugs are added from solutions prepared in methanol.

6. GC-MS positive control for stimulants: Add 200 ng/mL of amphetamine, fencamfamine, methadone, methamphetamine, phendimetrazine, and phenmetrazine to the negative control urine. All drugs are added from solutions prepared in methanol.

7. LC-MS/MS internal standard solution: Prepare a solution containing 0.1 µg/mL of d5-fentanyl and 20 µg/mL of d3-morphine-3β-D-glucuronide in methanol.

8. GC-MS internal standard for amphetamines: Prepare a 25 µg/mL solution of d5-amphetamine in ethyl acetate.

9. GC-MS phenazine internal standard: Prepare a 50 µg/mL solution of phenazine in isopropanol.

*2.4  Supplies*

1. High-performance liquid chromatography (HPLC) column: Reversed-phase biphenyl 100 Å, 2.6 μm particle size, 50 × 3 mm (Phenomenex).

2. Gas chromatography column: Ultra 1 100 % dimethylpolysiloxane, 17 m, 0.2 mm internal diameter, 0.33 μm film thickness (Agilent).

3. 15 mL glass round-bottom tubes with polytetrafluoroethylene (PTFE)-lined screw caps: These tubes are used for extraction of urine samples (GC-MS testing method).

4. 10 mL glass conical tubes with PTFE-lined screw caps: These tubes are used for derivatization of sample extracts (GC-MS testing method).

5. 10 mL glass round-bottom tubes with PTFE-lined screw caps: These tubes are used for sample dilution (LC-MS/MS testing method).

6. 2 mL auto sampler vials (Phenomenex).

7. 2 mL auto sampler vials with 0.3 mL inserts (Phenomenex).

8. Crimp caps for auto sampler vials with PTFE-lining (Phenomenex).

*2.5  Equipment*

1. LC-MS/MS system (AB Sciex API 4000 or API 4000 QTRAP) with electrospray interface and auto sampler.

2. GC-MS system (Agilent 6890/5972 or 5890/5972) operated in electron impact mode equipped with an auto sampler.

# 3  Methods

*3.1  LC-MS/MS Sample Preparation*

1. Place 0.1 mL of each urine sample into separate 10 mL round-bottom glass tubes. With every batch of samples, add 0.1 mL of negative control urine, positive control urine for stimulants, positive control urine for fentanyls, and positive control urine for narcotics into separate glass tubes (*see* **Note 2**).

2. To each tube add 44 μL of LC-MS/MS reagent mixture, cap, and gently swirl to mix (*see* **Note 3**).

3. Incubate for 1 h at 50 °C.

4. Transfer 100 μL to an auto sampler vial with a 0.3 mL insert and cap.

*3.2  LC-MS/MS Analysis*

1. Fill the mobile-phase reservoirs A and B with 0.1 % formic acid in water and 0.1 % formic acid in methanol, respectively.

2. Perform combined stimulants and narcotics acquisition for all samples and negative and positive control urine samples using the reversed-phase biphenyl HPLC column at a constant flow

rate of 0.6 mL per minute. The mobile phase gradient is initially 10 % methanol. The 10 % methanol is held for 0.5 min, ramped to 25 % during the next 1.5 min, ramped to 80 % methanol over the next 2.5 min, increased to 85 % and held for 1 min, then returned to 10 % methanol and held for 1.5 min.

3. Parameters for the electrospray ionization source and mass spectrometer are instrument-specific. For the AB Sciex API 4000 QTRAP instrument the parameters are as follows: curtain gas 20 PSI, collision gas medium, electrospray voltage 5500 V, sprayer temperature 450 °C, needle gas 30 PSI, heater gas 30 PSI, entrance potential 10 V, and declustering potential 35 V.

4. The mass spectrometer is operated in multiple reaction monitoring mode (MRM). Retention times and transitions (precursor/product ion pairs) for internal standards, narcotics and stimulants are shown in Tables 2 and 3. Between 1 and 3 transitions are monitored in MRM mode depending on the compound of interest.

## 3.3 LC-MS/MS Data Analysis

1. Integrate chromatograms for all MRMs and check all data for proper integration of the internal standards using Analyst software. Check for proper integration of all target compounds in the positive control urines.

2. Check for consistency of internal standard retention times between positive control urine, negative control urine and unknown samples. Internal standard retention times should be stable and within ±0.3 min of the retention times presented in Tables 2 and 3. Retention times for target compounds present in positive control urine are expected to be within ±0.5 min of the retention times presented in Tables 2 and 3.

3. The negative control urine must not contain any of the target compounds. The chromatograms should not have any integrated peaks within the expected retention time ranges for any of the monitored diagnostic transitions.

4. The peak heights of the internal standards d5-fentanyl and d3-morphine should be greater than 10,000 counts per second (cps). The peak heights of the narcotics in the control urine should be greater than 3000 cps. The peak heights of the stimulants in the control urine should be greater than 10,000 cps (*see* **Note 4**).

5. For unknown samples, a chromatographic peak in the window for a compound that has only one diagnostic transition that is within ±15 s of the expected retention time of the target compound is indicative of a positive result. In cases where multiple diagnostic transitions are monitored, the presence of a chromatographic peak for each of the transitions must be within ±15 s of the expected retention time of the target

**Table 2**
**Retention times and precursor/product ions (MRMs) for internal standards and narcotics by LC-MS/MS**

| Name | Retention time (min) | MRM 1 | MRM 2 |
| --- | --- | --- | --- |
| Internal standards | | | |
| d5-Fentanyl | 4.81 | 342/188 | |
| d3-Morphine | 1.53 | 289/201 | 289/153 |
| d3-Morphine glucuronide | 0.61 | 465/289 | 465/201 |
| Narcotics | | | |
| Alfentanil | 4.77 | 417/268 | 417/197 |
| Buprenorphine | 4.78 | 468/414 | 468/396 |
| Codeine | 3.92 | 300/215 | 300/165 |
| Fentanyl | 4.84 | 337/188 | 337/105 |
| Hydrocodone | 4.20 | 300/199 | 300/171 |
| Hydromorphone | 2.10 | 286/185 | 286/157 |
| Morphine | 1.62 | 286/201 | 286/153 |
| Norbuprenorphine | 5.45 | 414/187 | 414/414 |
| Norfentanyl | 4.28 | 233/84 | 233/177 |
| Noroxycodone | 3.76 | 302/284 | 302/187 |
| Oxycodone | 4.03 | 316/298 | 316/241 |
| Oxymorphone | 3.72 | 302/284 | 302/227 |
| Pethidine | 4.59 | 248/220 | 248/174 |
| Propoxyphene | 4.55 | 340/58 | 340/266 |
| Remifentanil | 4.14 | 377/317 | 377/345 |
| Sulfentanil | 4.90 | 387/238 | 387/111 |
| Tramadol | 3.96 | 264/58 | |

compound for a positive result (*see* **Note 5**). For compounds with several MRMs a comparison of the relative peak height for the ions in the unknown sample and the calibrator/positive control should be performed to increase the discriminating power of the testing procedure. If the relative abundance of the diagnostic ion(s) in the calibrator/positive control is between 1 and 25 % of the base peak, then the absolute abundance of the ion in the unknown sample must be ±5 %; if between 25 and 50 % then the relative abundance in the unknown sample must be ±20 %; if >50 % then the absolute abundance of the unknown sample must be within ±10 % [8].

**Table 3**
**Retention times and precursor/product ions (MRMs) for internal standards and stimulants by LC-MS/MS**

| Name | Retention time (min) | MRM 1 | MRM 2 | MRM 3 |
|---|---|---|---|---|
| Internal standards | | | | |
| d5-Fentanyl | 4.81 | 342/188 | | |
| d3-Morphine | 1.53 | 289/201 | 289/153 | |
| d3-Morphine glucuronide | 0.61 | 465/289 | 465/201 | |
| Stimulants | | | | |
| Benzoylecgonine | 4.40 | 290/168 | 290/105 | |
| Cropropamide | 4.80 | 241/196 | 241/100 | |
| Crotethamide | 4.65 | 227/182 | 227/154 | |
| Cyclozadone | 4.70 | 217/146 | 217/106 | |
| Dimethylamphetamine | 3.75 | 164/119 | 164/91 | |
| Dobutamine | 3.72 | 302/107 | 302/137 | 302/166 |
| Fenbutrazate | 4.90 | 368/191 | 368/119 | |
| Fencamine | 4.30 | 385/267 | 385/236 | |
| Heptaminol | 0.94 | 146/128 | 146/69 | |
| Isometheptene | 3.90 | 142/69 | 142/41 | |
| Methylhexaneamine | 2.90 | 116/99 | 116/57 | |
| Methylphenidate | 4.44 | 234/84 | 234/174 | |
| Norfenefrine | 0.70 | 154/136 | 154/119 | 154/91 |
| Octopamine | 0.70 | 154/91 | 154/119 | |
| Pentetrazol | 4.20 | 139/96 | 139/69 | |
| Phenpromethamine | 2.60 | 150/119 | 150/91 | |
| Ritalinic acid | 4.19 | 220/84 | 220/91 | 220/174 |
| Strychnine | 4.45 | 335/184 | 335/156 | |
| Tuaminoheptane | 2.90 | 116/57 | 116/43 | |

### 3.4 GC-MS Sample Preparation

1. Place 5 mL of urine into two separate 15 mL round-bottom tubes and cap the tubes with PTFE-lined screw caps. Duplicate aliquots of the calibrator, positive control, and negative control urine samples are included with each batch of unknown samples. Batches containing tubes 1 and 2 are treated as described below.

2. To tube 1, add 0.5 mL of a 9:1 hydrochloric acid:mercaptoacetic acid solution. Cap each tube and mix thoroughly.

3. Incubate tube 1 for 1 h at 95 °C.

4. Remove tube 1 from the heating block and allow to cool for 5 min at room temperature.

5. Add 0.6 mL of 12 N sodium hydroxide to both tubes 1 and 2, followed by 2 g of sodium bicarbonate:potassium carbonate solid buffer.

6. Add 2 mL of phenazine working solution to tubes 1 and 2, cap, and mix thoroughly for 5 min.

7. Centrifuge at $550 \times g$ for 10 min (*see* **Note 6**).

8. Add 30 µL of amphetamine internal standard to a 10 mL glass conical tube.

9. Quantitatively transfer 1 mL of the organic layer from tube 1 and 1 mL of the organic layer from the corresponding tube 2 into the 10 mL glass conical tube containing amphetamine internal standard.

10. Evaporate the solvent in the 10 mL glass conical tube to dryness under a gentle stream of nitrogen.

11. Add 100 µL of ethyl acetate, followed by 100 µL of trifluoro-acetic anhydride, the derivatizing reagent. Cap each tube and mix thoroughly.

12. Incubate for 15 min at 65 °C.

13. Evaporate the solvent and derivatization reagent at room temperature under a gentle stream of nitrogen (*see* **Note 7**).

14. Add 0.25 mL of ethyl acetate to each tube, mix thoroughly, and then transfer each sample to an auto sampler vial. Cap each vial.

### 3.5  GC-MS Analysis

1. Perform full-scan acquisition over $m/z$ range 50–550 for all samples including the negative and positive controls using an Ultra 1 100 % dimethylpolysiloxane column. Inject 1 µL of each sample in splitless mode with a splitless time of 18 s and an initial gas chromatography oven temperature of 110 °C. After 1 min the oven temperature is ramped to 300 °C during the next 7.6 min and then held at 300 °C for 3.4 min (*see* **Note 8**).

### 3.6  GC-MS Data Analysis

1. Process data for the calibrator first using the quantitative analysis system of the Agilent Chemstation software package (*see* **Note 9**). Ensure that all target compounds in the calibrator are correctly identified and have been assigned to a calibration level.

2. Process data for controls and all samples (*see* **Note 10**).

3. Check for consistency of internal standard retention times between calibrator, positive control urine, negative control urine, and samples. The retention time of phenazine should be within the range $5.25 \pm 2$ % minutes and the counts should be greater than 10,000. The retention time of d5-amphetamine should be within the range $2.93 \pm 2$ % minutes with $m/z$ ions

**Table 4**
**Retention times and ions for internal standards and narcotics by GC-MS**

| Name | Retention time (min) | m/z ion 1 | m/z ion 2 | m/z ion 3 |
|---|---|---|---|---|
| Internal standards | | | | |
| Phenazine | 5.25 | 180 | | |
| d5-Amphetamine | 2.93 | 144 | 123 | |
| Narcotics | | | | |
| Dextromoramide[a] | 10.30 | 100 | 128 | 265 |
| Norhydrocodone[a] | 8.62 | 241 | 381 | |
| Methadone[a] | 7.17 | 72 | 73 | 91 |
| Methadone metabolite (2-ethylidene-1, 5-dimethyl-3,3-diphenylpyrrolidine)[a] | 6.76 | 277 | 276 | 262 |
| Normethadone[a] | 6.99 | 58 | 224 | 165 |
| Pentazocine | 6.93 | 313 | 298 | |
| Tapentadol | 4.13 | 58 | 317 | |
| N-desmethyltapentadol | 5.28 | 141 | 203 | 330 |
| Tramadol | 4.92 | 188 | 159 | 173 |

[a]These compounds do not undergo derivatization

144, 123, and 92 all present. Retention times for target compounds present in calibrators and positive control urines should be within ±0.25 min of the retention times shown in Tables 4 and 5 (*see* **Note 11**).

4. The negative control urine must not contain any of the target compounds. Any target compounds identified by the quant software should be checked to verify that they are not the compound of interest.

5. The relative peak areas of cathine and ephedrine in the stimulant calibrator should be greater than one and three times that of phenazine, respectively. The sample report should indicate that the concentrations of cathine, ephedrine, pseudoephedrine, and caffeine are 2.5, 5, 5, and 3 µg/mL, respectively.

# 4   Notes

1. In addition to fentanyl, fentanyl derivatives are also prohibited (Fig. 1).

2. This procedure can also be carried out using a 96-well plate format with 0.1 mL sample aliquots by preparing a more

**Table 5**
**Retention times and ions for internal standards and stimulants by GC-MS**

| Name | Retention time (min) | m/z ion 1 | m/z ion 2 | m/z ion 3 |
|---|---|---|---|---|
| Internal standards | | | | |
| Phenazine | 5.25 | 180 | | |
| d5-Amphetamine | 2.93 | 144 | 123 | |
| Stimulants | | | | |
| 2-Methylamphetamine | 3.49 | 140 | 132 | 105 |
| 4-Methylamphetamine | 3.45 | 132 | 105 | 140 |
| Amfepramon[a] | 4.10 | 100 | 77 | 72 |
| Amiphenazole | 6.31 | 121 | 383 | |
| Amphetamine | 2.93 | 140 | 118 | 91 |
| Amphetaminil[a] | 5.47 | 132 | 105 | 133 |
| Benfluorex | 7.58 | 105 | 159 | 288 |
| Benzphetamine[a] | 5.90 | 91 | 148 | |
| Benzylpiperazine | 5.14 | 91 | 272 | 181 |
| β-Methylphenylethylamine | 3.07 | 105 | 118 | |
| Bromantane metabolite | 8.10 | 247 | 267 | |
| Caffeine[a] | 5.64 | 194 | 109 | |
| Carphedon artifact[a] | 5.90 | 104 | 200 | |
| Cathine | 3.15 | 140 | 230 | 203 |
| Chlorphentermine | 4.12 | 154 | 166 | |
| Clobenzorex | 6.80 | 125 | 127 | 264 |
| 4-Hydroxyclobenzorex | 7.03 | 125 | 127 | 264 |
| Clortermine | 3.99 | 154 | 166 | |
| Ephedrine | 3.56 | 154 | 110 | 244 |
| Ethamivan | 5.40 | 247 | 318 | |
| Ethylamphetamine | 3.85 | 168 | 140 | 118 |
| Etilefrine | 5.22 | 154 | 126 | |
| Famprofazone | 10.32 | 286 | 229 | 91 |
| Fencamfamine | 5.97 | 170 | 142 | |
| Fenetylline | 9.65 | 346 | 166 | |
| Fenfluramine | 3.74 | 168 | 159 | |

(continued)

**Table 5**
**(continued)**

| Name | Retention time (min) | *m/z* ion 1 | *m/z* ion 2 | *m/z* ion 3 |
|---|---|---|---|---|
| Fenproporex | 5.10 | 193 | 118 | |
| Furfenorex[a] | 5.05 | 81 | 138 | |
| Meclofenoxate[a] | 3.90 | 141 | 200 | 111 |
| Mefenorex | 5.30 | 216 | 218 | 140 |
| Mephedrone | 4.42 | 119 | 91 | 154 |
| Mephentermine | 3.86 | 168 | 110 | |
| Mesocarb (sydnonimine) | 7.56 | 91 | 119 | |
| Methamphetamine | 3.54 | 154 | 110 | 118 |
| *p*-Hydroxymethamphetamine | 4.40 | 154 | 110 | 230 |
| Methylenedioxyamphetamine | 4.70 | 135 | 162 | |
| Methylenedioxyethylamphetamine | 5.59 | 168 | 162 | 140 |
| Methylenedioxymethamphetamine | 5.30 | 154 | 162 | 135 |
| Methylephedrine[a] | 3.70 | 72 | 56 | |
| Methylone | 5.63 | 149 | 154 | 303 |
| Modafinil/Adrafinil | 8.53 | 167 | 165 | 152 |
| Modafinil/Adrafinil product[a] | 5.05 | 167 | 165 | 152 |
| *N*,α-Diethylphenylethylamine | 4.26 | 182 | 91 | 154 |
| *N*,β-Diethylphenylethylamine | 4.12 | 91 | 154 | |
| Nikethamide[a] | 4.25 | 106 | 78 | 177 |
| Desethylnikethamide | 3.29 | 106 | 78 | 107 |
| Norfenfluramine | 3.14 | 140 | 186 | 159 |
| Oxilofrine | 4.16 | 154 | 110 | 356 |
| Phendimetrazine[a] | 4.04 | 57 | 85 | |
| Phenmetrazine | 4.60 | 70 | 105 | |
| Phentermine | 3.05 | 154 | 132 | |
| Phenylephrine | 4.00 | 140 | 217 | |
| *p*-Hydroxyamphetamine | 3.95 | 140 | 230 | |
| Pipradol | 7.08 | 345 | 248 | |
| Prenylamine | 9.03 | 193 | 334 | 154 |
| Prenylamine metabolite (3,3-Diphenylpropylamine) | 6.34 | 167 | 152 | 307 |

(continued)

**Table 5**
**(continued)**

| Name | Retention time (min) | m/z ion 1 | m/z ion 2 | m/z ion 3 |
|---|---|---|---|---|
| Prolintane[a] | 4.82 | 126 | 127 | |
| Propylhexedrine | 3.50 | 154 | 110 | 182 |
| Pseudoephedrine | 3.87 | 154 | 110 | 155 |
| Selegiline[a] | 3.70 | 96 | 56 | 91 |
| Norselegiline | 4.10 | 178 | 118 | 91 |
| Sibutramine[a] | 5.85 | 114 | 72 | |
| N-desmethylsibutramine | 6.46 | 196 | 154 | 140 |
| N,N-didesmethylsibutramine | 5.88 | 165 | 137 | |
| Synephrine | 4.21 | 140 | 328 | |
| Trimetazidine | 7.26 | 181 | 166 | |

[a]These compounds do not undergo derivatization

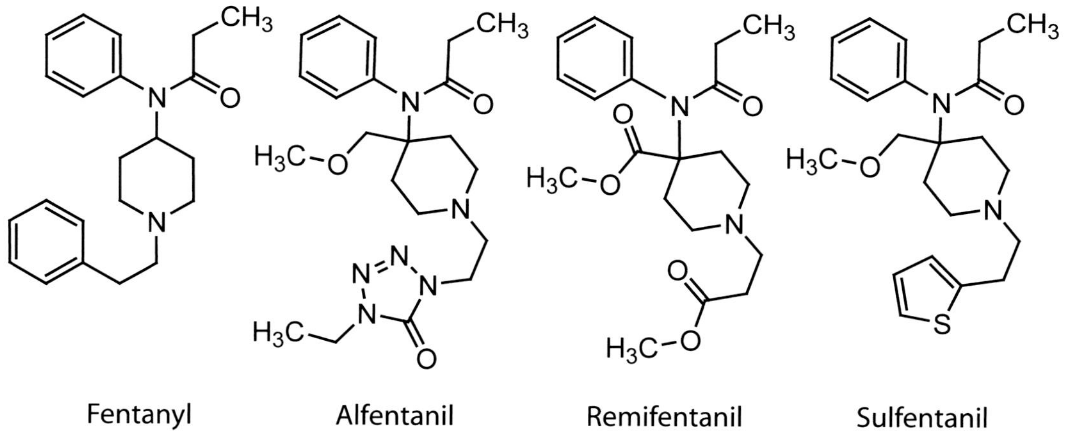

Fentanyl        Alfentanil        Remifentanil        Sulfentanil

**Fig. 1** In addition to fentanyl, all derivatives of fentanyl including alfentanil, remifentanil, and sufentanil are prohibited

concentrated internal standard solution for LC-MS/MS with d5-fentanyl and d3-morphine-3β-D-glucuronide at 4 μg/mL and 0.8 mg/mL in methanol, respectively. The LC-MS/MS reagent mixture is then prepared using 15 μL of this internal standard solution, 11.4 mL of ammonium bicarbonate buffer, and 0.6 mL of enzyme. Each well then receives 44 μL of this reagent mixture.

3. The LC-MS/MS reagent mixture is prepared daily. The recipe provided prepares enough reagent for five to ten unknown urine samples. The recipe can be scaled up or down depending upon the number of samples analyzed.

4. The minimum chromatographic peak heights and peak height ratios have been established specifically for the AB Sciex API 4000 QTRAP LC-MS/MS system using static interface settings for curtain gas, collision gas, electrospray voltage, sprayer temperature, needle gas, heater gas, and entrance and declustering potentials, and compound-specific settings for collision energy and exit potential. LC-MS/MS instrument parameters and interface optimization will dramatically alter both the relative and absolute responses for the diagnostic ions being monitored by this method. If different instrumentation is used to detect these compounds validation studies will need to be performed to establish criteria for minimum chromatographic peak heights and peak ratios.

5. An example showing selected ion chromatographs for ritalinic acid in the positive control, negative control, and a positive athlete's sample are shown in Fig. 2.

6. Only the organic layer should be transferred. If there is an emulsion layer that cannot be eliminated by further addition of 0.5 g salt followed by centrifugation, do not transfer the emulsion since it will drastically increase the drying time.

7. The evaporation time must be closely monitored so that volatile target compounds are not lost after removal of the solvent.

8. A solvent delay time of 1.2 min should be used to increase the lifetime of the filament and electron multiplier.

9. Data analysis for this procedure is usually performed using Quant within Agilent Chemstation G1701EA Revision E.02.00 using the RTEINT integrator and extracting signals for target and qualifier $m/z$ ions over a 0.5 min time range centered on the expected retention times listed in Tables 4 and 5. Background subtraction is performed using the "low first and last" setting with identification requiring all qualifiers to be met and the best retention time match used to select between multiple hits.

10. Depending on the compound being detected by the GC-MS testing method the unaltered parent compound, a metabolite(s), or both can be detected in the urine. Following the administration of clobenzorex, parent compound is not detected, whereas the main metabolite amphetamine and to a lesser degree the characteristic metabolite 4-hydroxyclobenzorex are detected, as shown in Fig. 3.

11. Compounds marked with an asterisk in Tables 4 and 5 are monitored as the underivatized compound due to the lack of

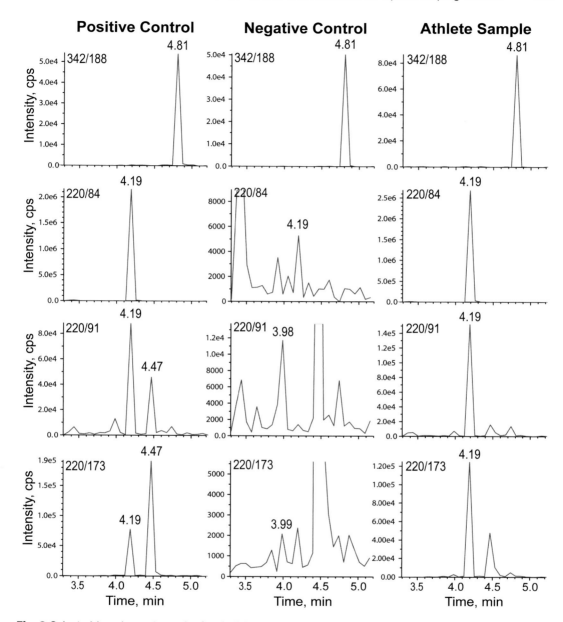

**Fig. 2** Selected ion chromatographs for ritalinic acid in a positive control urine (*left*), negative control urine (*middle*) and an athlete's positive urine sample (*right*). The *top panels* are the MRM *m/z* 342 → *m/z* 188 for the internal standard d5-fentanyl. Ritalinic acid MRM *m/z* 220 → *m/z* 84, MRM *m/z* 220 → *m/z* 91 and MRM *m/z* 220 → *m/z* 173 are shown in the *bottom three panels*. The retention time of ritalinic acid is 4.19 min in the positive control

a functional group that can undergo acylation. Derivatized compounds often yield spectra with larger *m/z* fragments that are in high abundance. This is illustrated in the spectra of amphetamine following derivatization with trifluoroacetic anhydride (Fig. 4).

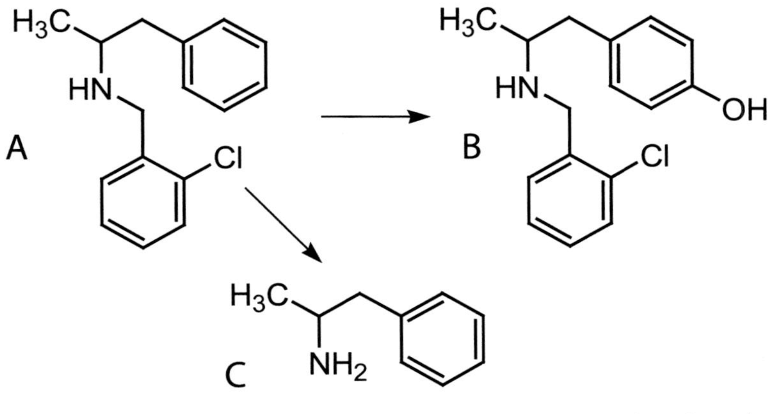

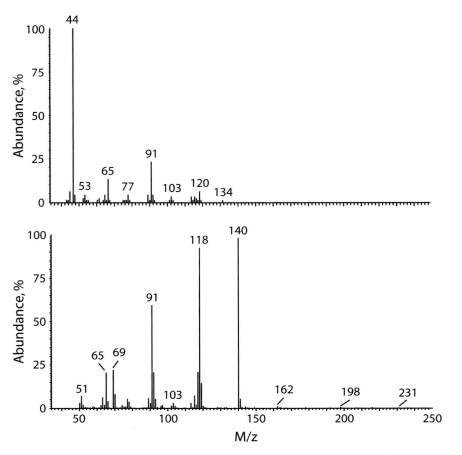

**Fig. 3** Only trace amounts of unchanged drug is present in urine after administration of the stimulant clobenzorex (A). The major metabolites are 4-hydroxyclobenzorex (B) and amphetamine (C)

**Fig. 4** Mass spectra of amphetamine before (*top panel*) and after derivatization with trifluoroacetic anhydride to produce *N*-trifluoroacetylamphetamine (*bottom panel*)

## References

1. The World Anti-Doping Code. The 2015 prohibited list. https://wada-main-prod.s3.amazonaws.com/resources/files/wada-2015-prohibited-list-en.pdf. Accessed 17 Feb 2015

2. Minimum required performance levels for detection and identification of non-threshold substances. https://wada-main-prod.s3.amazonaws.com/resources/files/wada_td2015mrpl_minimum_required_perf_levels_en.pdf. Accessed 28 Sep 2015

3. Solans A, Carnicero M, de la Torre R, Segura J (1995) Comprehensive screening procedure for detection of stimulants, narcotics, adrenergic drugs, and their metabolites in human urine. J Anal Toxicol 19:104–114

4. Hemmersbach P, de la Torre R (1996) Stimulants, narcotics, and beta-blockers: 25 years of development in analytical techniques for doping control. J Chromatogr B 687:221–238

5. Segura J, Ventura R, Jurado C (1998) Derivatization procedures for gas chromatographic-mass spectrometric determination of xenobiotics in biological samples, with special attention to drugs of abuse and doping agents. J Chromatogr B 713:61–90

6. Guddat S, Solymos E, Orlovius A, Thomas A, Sigmund G, Geyer H, Thevis M, Schanzer W (2011) High-throughput screening for various classes of doping agents using a new 'dilute-and-shoot' liquid chromatography-tandem mass spectrometry multi-target approach. Drug Test Anal 3:836–850

7. Deventer K, Pozo OJ, Van Eenoo P, Delbeke FT (2009) Qualitative detection of diuretics and acidic metabolites of other doping agents in human urine by high-performance liquid chromatography-tandem mass spectrometry. Comparison between liquid-liquid extraction and direct injection. J Chromatogr B 1216: 5819–5827

8. Identification criteria for qualitative assays incorporating column chromatography and mass spectrometry. https://wada-main-prod.s3.amazonaws.com/resources/files/wada_td2015idcr_minimum_criteria_chromato-mass_spectro_conf_en.pdf. Accessed 28 Sep 2015

# Chapter 27

## Quantification of Tricyclic Antidepressants in Serum Using Liquid Chromatography Electrospray Tandem Mass Spectrometry (HPLC-ESI-MS/MS)

### Christopher A. Crutchfield, Autumn R. Breaud, and William A. Clarke

### Abstract

Tricyclic antidepressants (TCA) are used to treat major depressive disorder and other psychological conditions. The efficacy of these drugs is tied to a narrow therapeutic window. Inappropriately high drug concentrations can result in serious side effects such as hypotension, tachycardia, or coma. As a result, concentrations of tricyclic antidepressants are routinely monitored to ensure compliance and to prevent adverse side effects by dose adjustments. We describe a method for the determination of concentrations of amitriptyline, desipramine, imipramine, and nortriptyline in human serum using high-performance liquid chromatography coupled to a tandem mass spectrometer with electrospray ionization (HPLC-ESI-MS/MS). The method is rapid, requiring only 3.5 min per analysis. The method requires 100 μL of serum. Concentrations of each TCA were quantified by a calibration curve relating the peak area ratio of each TCA analyte to a deuterated internal standard (amitriptyline-D3, desipramine-D3, imipramine-D3, and nortriptyline-D3). The method was linear from ~70 ng/mL to ~1000 ng/mL for all TCAs, with imprecision $\leq 12$ %.

**Key words** Tricyclic antidepressants, Depression, Tandem mass spectrometer, Amitriptyline, Desipramine, Imipramine, Nortriptyline

## 1 Introduction

Tricyclic antidepressants (TCA) are a class of drugs used to treat major depressive disorder and other psychiatric conditions. Typically other pharmacological agents are used prior to TCA due to the increased risk of side effects of TCA use, including hypotension, tachycardia, coma, respiratory depression, and in cases of overdose, death. However, if drug levels are too low, the patient may not receive the pharmacological benefit. As a result, TCA are a good candidate for therapeutic drug monitoring. Immunoassay-based measurement of tricyclic antidepressants exists, but is susceptible to interference [1–3]. This method [4] is much more rapid than previous LC-MS/MS-based methods for TCA

Uttam Garg (ed.), *Clinical Applications of Mass Spectrometry in Drug Analysis: Methods and Protocols*, Methods in Molecular Biology, vol. 1383, DOI 10.1007/978-1-4939-3252-8_27, © Springer Science+Business Media New York 2016

quantification, requiring only 3.5 min per analysis compared to 24 min [5] and 20 min [6]. This method enables rapid and reliable TCA quantification in serum.

# 2    Materials

## 2.1    Sample (Human Serum)

Serum separator tubes are unacceptable. Samples are stable for 2 weeks at 4 °C.

## 2.2    Solvents and Reagents

1. Mobile Phase A, 0.1 % (v/v) formic acid in HPLC-grade water, stable for 1 month at room temperature, 18–24 °C.

2. Mobile Phase B, 0.1 % (v/v) formic acid in HPLC-grade acetonitrile, stable for 1 month at room temperature, 18–24 °C.

3. Human drug-free pooled normal serum.

## 2.3    Internal Standards and Standards

1. Primary standards: (1 mg/mL amitriptyline, desipramine, imipramine, and nortriptyline) (Cerilliant).

2. Primary internal standards: (100 µg/mL amitriptyline-$D_3$, desipramine-$D_3$, imipramine-$D_3$, and nortriptyline-$D_3$) (Cerilliant).

3. Primary Standard Working Solutions: amitriptyline, desipramine, imipramine, and nortriptyline are pooled into working solutions at levels of 10 µg/mL, 4 µg/mL, and 400 ng/mL in methanol.

4. I.S. Working Solution/Extraction Solution (48 ng/mL amitriptyline, desipramine, imipramine, and nortriptyline in methanol): Add 240 µL of 100 µg/mL from each amitriptyline-$D_3$, desipramine-$D_3$, imipramine-$D_3$, and nortriptyline-$D_3$ stock solution to a class A 500 mL volumetric flask, fill to level with methanol, and mix. Stable for 3 months at 4 °C.

## 2.4    Calibrators and Controls

1. Calibrators: Prepare calibrators 1–6 (**Note 1**) by diluting working stock solutions with drug-free normal human serum in 10 mL class A volumetric flasks (Table 1).

2. Controls: Bio-Rad Lyphocheck Benzo/TCA Control Set A Control Level 1 and Bio-Rad Lyphocheck Benzo/TCA Control Set A Control Level 1 (Bio-Rad).

## 2.5    Analytical Equipment and Supplies

1. Finnegan Surveyor MS Pump Plus with a Finnegan Surveyor Autosampler Plus coupled to a TSQ Quantum Access tandem mass spectrometer (Thermo Fisher Scientific).

2. Analytical column: Thermo Scientific Hypersil Gold C-18, 2.1 × 50 mm, particle size 3 µm (Thermo Fisher Scientific).

3. 1.8 mL glass HPLC vials.

4. 1.5 mL polypropylene microcentrifuge tubes.

**Table 1**
**Preparation of calibrators**

| Calibrator | Working stock concentration (ng/mL) | Working stock volume (µL) | Final volume (mL) | Final concentration (ng/mL) |
|---|---|---|---|---|
| 1 | 400 | 375 | 10 | 15 |
| 2 | 400 | 1000 | 10 | 40 |
| 3 | 4000 | 250 | 10 | 100 |
| 4 | 4000 | 625 | 10 | 250 |
| 5 | 10,000 | 500 | 10 | 500 |
| 6 | 10,000 | 1000 | 10 | 1000 |

# 3   Methods

### 3.1   Stepwise Procedure

1. To a labeled 1.5 mL polypropylene centrifuge tube, pipette 100 µL of serum (calibrator, control, or unknown sample).
2. Add 500 µL of extraction solution.
3. Cap and vortex for 20 s.
4. Centrifuge for 5 min at $18,000 \times g$.
5. Dilute 450 µL supernatant 1:1 with HPLC-grade water in a labeled 1.8 mL glass vial.
6. Cap and vortex briefly.
7. Please vials into autosampler.
8. Inject 10 µL and analyze.

### 3.2   Sample Analysis

1. Instrumental operating parameters are given in Table 2.
2. Data are analyzed using LCQuan (Thermo Scientific).
3. Standard curves are generated based on linear regression with $1/x^2$ weighting of the analyte/internal standard peak-area ratio relative to the nominal analyte concentration. Correlation coefficients are typically $r^2 > 0.995$.
4. Acceptability criteria are based on modified Westgard rules.
   (a) Value exceeds 3 sd of established mean.
   (b) 2 values in a row exceed 2 sd.
   (c) 6 or more values in a row trend positive or negative bias.
5. Imprecision is typically $\leq 12$ % at all QC levels.
6. Representative chromatograms and mass transitions of TCAs are shown in Fig. 1. (**Note 2**).

**Table 2**
**HPLC-MS/MS operation conditions**

| HPLC | | |
|---|---|---|
| **Time (min)** | **Flow rate (µL/min)** | **Mobile phase A (%)** |
| 0 | 600 | 80 |
| 0.5 | 600 | 80 |
| 0.55 | 400 | 80 |
| 1 | 400 | 50 |
| 2 | 400 | 0 |
| 2.5 | 400 | 0 |
| 2.55 | 400 | 80 |
| 2.6 | 600 | 80 |
| 3.5 | 600 | 80 |

| MS/MS Tune Settings | |
|---|---|
| **Parameter** | **Value** |
| Spray voltage (V) | 4900 |
| Sheath gas | 30 |
| Aux gas | 10 |
| Capillary temperature (°C) | 270 |

| Precursor and product ions for tricyclic antidepressants | | | | | |
|---|---|---|---|---|---|
| **Compound** | **Precursor** | **Primary Product** | **Primary CE (eV)** | **Secondary Product** | **Secondary CE (eV)** |
| Amitriptyline | 278.2 | 233.1 | 18 | 202.1 | 55 |
| Desipramine | 267.1 | 72.3 | 14 | 193.1 | 36 |
| Imipramine | 281 | 86.2 | 15 | 193.1 | 42 |
| Nortriptyline | 264.2 | 202.1 | 58 | 233.1 | 15 |
| Amitriptyline-D3 | 281.2 | 233.1 | 16 | 78 | |
| Desipramine-D3 | 270.2 | 75.3 | 16 | 73 | |
| Imipramine-D3 | 284.2 | 89.3 | 16 | 60 | |
| Nortriptyline-D3 | 267.2 | 233.1 | 14 | 80 | |

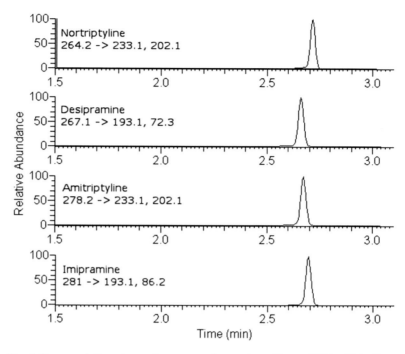

**Fig. 1** Representative chromatograms and mass transitions of TCAs (**Note 3**)

## 4   Notes

1. Individual sets of calibrators 1–6 may be pre-aliquoted and frozen until use. These materials are stable for 1 year when stored unopened at −80 °C.

2. Matrix effects were evaluated using post-column infusion as well as comparison of spiked sera and spiked solvent. Matrix effects were <12 % for all analytes.

3. Retention time will be system-specific, but using this method all analytes co-eluted at ~2.6 min. Desipramine and imipramine share a secondary product. They do exhibit transition cross-talk.

### References

1. McKay CA, Wu AHB (2002) Quetiapine (SeroquelTM by astrazeneca) interference with tricyclic antidepressant immunoassays. J Toxicol Clin Toxicol 40:661

2. Saidinejad M, Law T, Ewald MB (2007) Interference by carbamazepine and oxcarbazepine with serum- and urine-screening assays for tricyclic antidepressants. Pediatrics 120:E504–E509

3. Song D, Chin W (2009) Analytical interference of carbamazepine on the Abbott TDx and Abbott Axsym tricyclic antidepressant assays. Pathology (Phila) 41:688–689

4. Breaud AR, Harlan R, Di Bussolo JM, McMillin GA, Clarke W (2010) A rapid and fully-automated method for the quantitation of tricyclic antidepressants in serum using turbulent-flow liquid chromatography–tandem mass spectrometry. Clin Chim Acta 411: 825–832

5. Titier K, Castaing N, Le-Déodic M, Delphine L-b, Moore N, Molimard M (2007) Quantification of tricyclic antidepressants and monoamine oxidase inhibitors by high-performance liquid chromatography-tandem mass spectrometry in whole blood. J Anal Toxicol 31:200–207

6. Alves C, Santos-Neto AJ, Fernandes C, Rodrigues JC, Lanças FM (2007) Analysis of tricyclic antidepressant drugs in plasma by means of solid-phase microextraction-liquid chromatography-mass spectrometry. J Mass Spectrom 42:1342–1347

# INDEX

Uttam Garg (ed.), *Clinical Applications of Mass Spectrometry in Drug Analysis: Methods and Protocols*, Methods in Molecular Biology, vol. 1383, DOI 10.1007/978-1-4939-3252-8, © Springer Science+Business Media New York 2016